Fractals in Chemistry

Andrew Harrison

Department of Chemistry, University of Edinburgh

OXFORD   NEW YORK   TOKYO
OXFORD UNIVERSITY PRESS
1995

Oxford University Press, Walton Street, Oxford OX2 6DP

Oxford New York
Athens Auckland Bangkok Bombay
Calcutta Cape Town Dar es Salaam Delhi
Florence Hong Kong Istanbul Karachi
Kuala Lumpur Madras Madrid Melbourne
Mexico City Nairobi Paris Singapore
Taipei Tokyo Toronto

and associated companies in
Berlin Ibadan

Oxford is a trade mark of Oxford University Press

Published in the United States
by Oxford University Press Inc., New York

A catalogue record for this book is available from the British Library

Library of Congress Cataloging in Publication Data
Harrison, Andrew.
Fractals in Chemistry / Andrew Harrison.
(Oxford chemistry primers ; 22)
1. Chemistry, Physical and theoretical--Mathematics.
2. Fractals. I. Title. II. Series.
QD455.3.M3H38 1995 541'.01'51474--dc20 94-46223

ISBN 0 19 855768 X (Hbk)
ISBN 0 19 855767 1 (Pbk)

Typeset by the author

Printed in Great Britain by
The Bath Press, Avon.

Series Editor's Foreword

Oxford Chemistry Primers are designed to provide clear and concise introductions to a wide range of topics that may be encountered by chemistry students as they progress from the freshman stage through to graduation. The Physical Chemistry series will contain books easily recognized as relating to established fundamental core material that all chemists will need to know, as well as books reflecting new directions and research trends in the subject, thereby anticipating (and perhaps encouraging) the evolution of modern undergraduate courses.

In this Physical Chemistry Primer, Andrew Harrison has produced a stimulating, easy-to-read introduction to the use of *Fractals in Chemistry*. The Primer will interest all students (and their mentors) who wish to appreciate the importance of fractals in a broad diversity of chemical phenomena.

Richard G.Compton
Physical Chemistry Laboratory, University of Oxford

Preface

The rugged west coast of Scotland, dendritic forms of minerals, and the fern-like patterns made by frost on a cold window all have irregular shapes whose outline has a length that appears to increase as it is scrutinized in greater and greater detail. The fractal dimension is a concept that allows us to quantify the perimeter or surface of such objects, and defines a new form of symmetry that bestows some order on the apparent chaos. The elegance of the concept and the beauty of some of the objects it describes has led to overexposure in recent years; nevertheless, the study of fractal objects contains important lessons for chemists who work with porous solids, rough surfaces, polymers, and colloids. In this book I introduce a single, simple model that unifies the growth of such objects, present experiments to produce fractal objects in a laboratory, and describe techniques to measure their properties. I show how the fractal dimension may control physical and chemical reactions in and on porous solids, and discuss the consequences this has for heterogeneous chemistry.

I am grateful for the help given by many people in many ways while I wrote and illustrated this book. Those who freely offered photographs and figures are too numerous to cite here individually and are listed overleaf. I am particularly grateful to Monica Price of the University Museum, Oxford for her help with minerals, to Len Cumming in Edinburgh for much of the photography, and to Paul Bourke of the University of Auckland for his help with his box-counting program. I thank Andrew Wills of Edinburgh University who tested many of the experiments described in Chapter 2.

The staff of Oxford University Press have listened to all my excuses and answered all my queries with great patience, and Richard Compton has offered guidance and encouragement when I needed it. Finally, and most importantly, I thank Alison and my parents for all their support. I dedicate this book to all those who have taught me.

Edinburgh   A.H.
May, 1994

Acknowledgements

I am very grateful to David Avnir, Henri Van Damme, Raoul Kopelman, Mitsugu Matsushita, Arne Skjeltorp, Sir John Meurig Thomas at the Royal Institution, and David Weitz for sending me photographs and diagrams for reproduction here; I also wish to thank Andrew Dougherty, Paul Meakin, and Dale Schaefer for their help in copying or adapting their figures.

Permission to reproduce Fig. 1.7 was given by *Nature,* and the Physical Society of Japan provided permission to reproduce Fig. 2.17. Fig. 2.3 has been reprinted from *Solid State Communications*, Vol. 60, J.H. Kaufman, O.R. Melroy, F.F. Abraham and A.I. Nazzal, Growth Instability in Diffusion Controlled Polymerisation, 757–761, 1986 and Fig. 4.9 has been reprinted from *Electrochimica Acta*, Vol. 34, T. Pajkossy and L. Nyikos, Diffusion to Fractal Surfaces –II. Verification of Theory, 171–179, 1989 with permission from Elsevier Science Ltd., Pergamon Imprint, The Boulevard, Langford Lane, Kidlington OX5 1GB, UK. Figs. 2.1(a), 2.5(b), 2.6(b), and 2.16 (a–c) are reproduced by permission of John Wiley and Sons Limited from *The Fractal Approach to Heterogeneous Chemistry*, edited by D. Avnir (©1989, John Wiley and Sons Limited).

Contents

1 Fractal forms

1.1 Scale and symmetry in chemistry

One of the great triumphs of the physical sciences this century has been the development of techniques to study the size and shape of objects at an atomic or molecular level, and the application of these techniques to help us understand the way in which atoms and molecules interact. If such objects are condensed to form crystalline solids, translational symmetry allows us to predict the spatial relationships between any atoms in the solid. Without translational symmetry, as is the case for a glass, we have to be satisfied with a knowledge of the local geometry for a particular atom at short length-scales (≈ 0.1 nm), a poorly defined distribution of atoms at moderate length-scales (0.2–1 nm), and bulk values of properties such as density at larger length-scales. A chemist might regard this degree of detail as quite sufficient for dealing with chemical problems. For example, he or she might be content to describe a porous solid in terms of its chemical composition, density, and perhaps a surface area. In the field of heterogeneous catalysis, such areas are of great importance, and chemists have developed techniques to quantify them – commonly by measuring the number of gas molecules of a particular size needed for monolayer coverage. Armed with this knowledge, one might describe the area in terms of football pitches per kilogram – or more formally as m^2kg^{-1} – and use this information to predict the coverage expected for gas molecules of a different size. We shall see below that such predictions are likely to be wrong and we will consider how an irregular object may have a particular type of symmetry that allows us to the predict spatial distribution of matter over wide ranges of length-scale.

The area of a rough surface

How do we quantify the area and degree of irregularity of a rough surface? Conventionally the area may be measured in one of two ways. It may be tiled with a suitable shape of known area σ or, if it has a shape or outline whose area depends in a known manner on its linear dimensions such as length of edge or radius, we may measure these dimensions and calculate the area. Both approaches are illustrated for a square object in Fig. 1.1.

The area A of the surface is either the sum of the tile areas or the product of the length of the sides. In the case where the tiles have edges of unit length λ, and the edge of the complete square has a length L such that it is an integral multiple $n(\lambda)$ of λ, A is given either by

$$A = L^2 = (n(\lambda)\,\lambda)^2 = n(\sigma)\,\lambda^2 \qquad (1.1)$$

or

$$A = n(\sigma)\,\sigma = (n(\lambda)\,\lambda)^2 \tag{1.2}$$

where

$$n(\sigma) = (n(\lambda))^2 \tag{1.3}$$

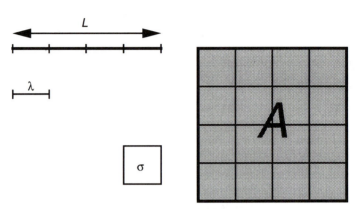

Fig. 1.1 Relations between the edge and area of tiles and planar objects.

In both cases the units of the area are λ^D, where the dimension D is the integer 2. We shall soon see that we ought to be careful about the way we use the word 'dimension'. At present we will use an intuitive definition which takes the dimension of an object to be the minimum number of coordinates necessary to specify the position of a point contained in that object. For example, we need only specify the distance from the origin to define the position of a point on a line but two coordinates are required to specify a position in a plane. It is clear that a point has dimension zero.

A chemist might estimate the area of a surface with the aid of molecular tiles which adhere either relatively weakly through Van der Waals forces, or strongly through the formation of a chemical bond. These processes are known as *physi*sorption and *chemi*sorption respectively and have energies of the order of -20kJ mol^{-1} and -200 kJ mol^{-1}; they will be considered in greater detail in Chapter 4. For the moment we will ignore the microscopic nature of the adsorption and assume that gas molecules bind to equivalent empty sites until the surface is uniformly covered with a layer one molecule thick.

If we know the mean area σ taken up by one molecule, as well as the number of molecules $n(\sigma)$ that are adsorbed, the surface area of adsorbate is given by

$$A = n(\sigma)\,\sigma \tag{1.4}$$

Estimates of σ may be obtained from the density of liquefied gases or from bond lengths and molecular geometry. Traditionally they were also obtained by assuming that nitrogen had a standard value for σ of 16.2 Å^2 per molecule and calculating coverage in terms of this unit area. As we shall see, this last method is flawed. If the surface of the adsorbate is perfectly smooth, then we expect the estimate of the area to be independent of the size of the probe molecule. However, if the surface is very rough or contains pores that are small compared with σ we expect that as σ increases, less of the surface is accessible, as illustrated in Fig. 1.2.

Rough or porous solids are common in heterogeneous chemistry because they provide a large active area per kilogram. The active surface may either be provided by the pure porous solid, or by some other compound which is supported by the porous solid. Many of the materials that are porous at molecular length-scales are solids that are built from molecules or collections of molecules as open frameworks. One of the most widely used examples of this type of material is provided by the zeolites, which are aluminosilicate compounds of general formula $M_{x/n}[(AlO_2)_x (SiO_2)_y].wH_2O$ where M is commonly a group 1 or 2 element of valence n. They may be found as minerals or prepared from hot, aqueous solutions containing aluminate and silicate ions and OH^-. Polymerization of the oxoanions over a period of hours or days leads to the formation of crystals with a variety of cage structures, depending on the reaction conditions.

Fig. 1.3 shows the zeolite ZSM-5 at various length-scales but with the same orientation. At the top left of the Figure there is a representation of the aluminosilicate skeleton with the vertices representing the various atoms. Below it there is a transmission electron micrograph that shows how the channels in the structure repeat in a regular fashion, and the regular array of apertures through which molecules must pass before they can enter the solid; at lower magnification we observe smooth crystals, shown on the right.

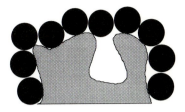

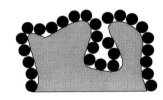

Fig. 1.2 Cross-section of a rough surface covered with probe molecules of two different sizes.

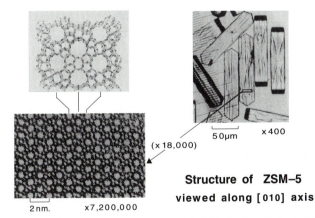

(×18,000)

50 μm   ×400

2 nm.   ×7,200,000

Structure of ZSM–5
viewed along [010] axis

Fig. 1.3 The photograph at the top left shows a model of the aluminosilicate skeleton of ZSM-5. Below it there is a transmission electron micrograph that shows how this cell repeats to give a regular porous network at molecular length-scales while at longer length-scales, crystals of this material appear smooth, as shown to the right.

The size of the aperture may be controlled through the choice of the cation M which allows us to devise molecular sieves with different meshes. In the case of zeolite A, the apertures are 3 Å, 4 Å, and 4.9 Å for the K, Na, and Ca salts respectively; these materials are commonly known as the standard laboratory molecular sieves 3A, 4A, and 5A.

Porous solids with far less regularity may be produced by rapid polymerization of silicate ions in either acidic or basic solutions. Initial polymerization produces small, spherical particles whose composition is close to SiO_2 and which are bound through Si—O—Si bridges, with Si—OH or Si—O$^-$ bonds on the surface. In acidic solution these may coalesce to produce a network of fine particles that encloses some of the solvent. This is called a gel. In basic solution, the particles are anionic and are stabilized against aggregation by electrostatic repulsion. They continue to grow in a manner that may be controlled through the pH, concentration of silicate, and temperature. If the pH is then lowered, or the ionic strength of the solution increased through the addition of ionic salts, the particles may approach one another more closely and bind through Van der Waals forces. The gel that is then produced has much larger pores than that produced under basic conditions. The control of the pore size through pH is depicted schematically in Fig. 1.4.

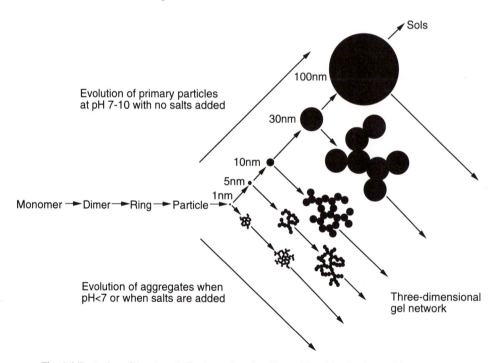

Fig. 1.4 Illustration of the steps in the formation of a silica gel in acid or basic conditions.

The solvent may be removed by heating the solid to produce a xerogel (*xeros* is the ancient Greek word for 'dry') whose cavities depend both on the characteristics of the primary particles and on how they bind. In Fig. 1.5. we show a representation of a micrograph of the porous surface provided by such a material, and in Fig. 1.6. we compare the distribution of pore sizes for a silica gel that has been grown so that the mean pore size is 60 Å, and a form of zeolite A in which the cation M is Na⁺.

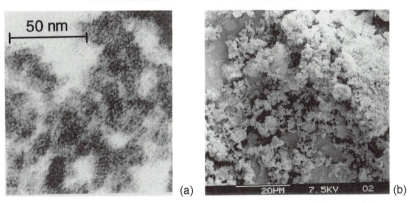

(a)   (b)

Fig. 1.5 Representations of electron micrographs of silica gels at (a) high resolution, depicting the primary particles and (b) lower resolution, showing a highly porous surface.

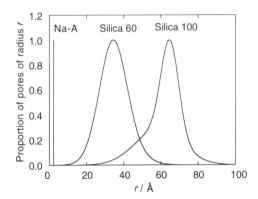

Fig. 1.6 Distribution of pore sizes for zeolite A in which M is Na and the silica gels silica 60 and silica 100

Porous solids may also be prepared by the erosion of denser materials. One good example of this method is based on borosilicate glasses. If a melt of sodium borate and sodium silicate is cooled below a certain temperature, the liquids become immiscible so that when the melt is quenched to form a glass, it contains segregated regions rich and poor in the borate. The area of the interface, and hence the mean width of these regions, may be finely controlled through the thermal history of the sample. If the glass is crushed and heated in 6N aqueous HCL the sodium borate is leached out, leaving a porous silicate glass. The distribution of the pore size is commonly much narrower than for silica gels and such materials are often called controlled-pore glasses (Fig. 1.7) and denoted cpg *n*, where *n* is the mean pore diameter in Å.

In Fig. 1.8 we show experimental data for monolayer coverage of silica 60 and cpg 75. The probe molecule is an alkane and we see that as its size increases, $n(\sigma)$ decreases in a linear fashion when plotted on double logarithmic axes.

Fig. 1.7 Electron micrograph of a controlled-pore glass.

The following empirical relationship may be extracted from the graph:

$$\log n(\sigma) = -\alpha \log\sigma + \beta \qquad (1.5)$$

or

$$n(\sigma) = \beta\, \sigma^{-\alpha} \qquad (1.6)$$

where α and β are constants for this surface and adsorbate. If eqn 1.4 is invoked, the area A as a function of σ becomes

$$A(\sigma) = \beta\, \sigma^{1-\alpha} \qquad (1.7)$$

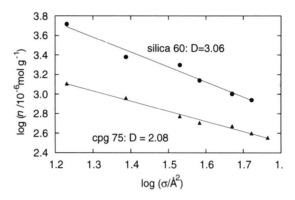

Fig. 1.8 The dependence of the surface coverage for silica 60 and cpg 75 on the cross-sectional area of *n*-alkanes.

We may also express this in terms of a linear dimension λ, which we could take to be the average radius of the probe molecule such that

$$\sigma = k\, \lambda^2 \qquad (1.8)$$

where k depends of the shape of the probe. Equations 1.5 to 1.7 now become

$$\log n(\lambda) = -2\alpha \log\lambda + \beta' \qquad (1.9)$$

$$n(\lambda) = \beta'\, \lambda^{-2\alpha} \qquad (1.10)$$

$$A(\sigma) = k\beta'\, \alpha^{2-2\alpha} = \beta''\, \alpha^{2-2\alpha} \qquad (1.11)$$

For a flat surface $\alpha = 1$ and the measure of the area is independent of σ or λ, whereas in these cases α takes values *greater* than 1. As the probe molecule decreases in size, so a greater proportion of the surface becomes accessible; the gradient α in Fig. 1.8 and eqns 1.5–1.7 is related to the degree to which new features are unearthed as σ or λ are reduced. It is important to note that α is not merely an index of porosity. This may be appreciated by considering the behaviour of zeolite 4A. If the probe molecule is larger than 4 Å it cannot enter the solid, and the material appears smooth. A plot of $\log(A)$ against $\log(\sigma)$ would then be flat until σ was sufficiently small to enter the pores, whereupon the line would take a leap upwards. As σ is further reduced, the molecule regards the surface provided by the adsorbate σ as smooth once more, and the graph settles down to a second flat region. The constant β in eqns 1.5 to 1.7 and β' in 1.9 and 1.10 reflects the porosity of the surface per unit mass. It is sometimes referred to as the *lacunarity* of the material, derived from the Latin word *lacuna* for gap and provides one index of texture. The more 'holey' the surface, the higher the value of the prefactors β or β'.

Before we look more closely at the interpretation of Fig. 1.8 and the nature of α, let us look at a related problem that concerns the length rather than the area of an irregular boundary.

Around the rugged rocks...

The question 'How long is the coast of Britain?' has puzzled thinkers in various forms for millennia. The boundary in question could be a meandering river, or the outline of a soot particle. What matters is that it continues to have an irregular form as it is observed in greater and greater detail.

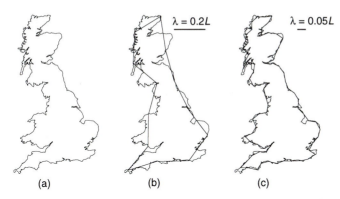

Fig. 1.9 The length of the coast of the largest British Island may be estimated with a set of dividers. We display the results of such measurements with dividers of length $\lambda = 0.2$ and $0.05L$, where L is the maximum width of the island.

In Fig. 1.9(a) we depict the coastline of the largest island of the British Isles with a line whose width is approximately 1.4 km to the scale of the figure. How do we measure the length of the boundary that this line represents? One straightforward, if somewhat laborious way would be to strap a pedometer to someone with time on their hands, and induce them to walk around the coastal path. An arbitrary decision could be made to cross rivers by the bridge closest to the mouth, or, where the river is too small or remote to merit a bridge, to step across it when it is narrower than 1 m. A year later we might obtain a measure of the length and this would be of the order of 10^4 km. The pedantic reader might then point out that the coastal path omits many of the finer features of the coastline, and that the walker should follow every boulder along the shore, taking care to step around it with a smaller stride. If we continue this argument much further and require that the length be measured around every rock, pebble, and grain of sand, the extent of the coastline would appear to be limited only by the discreet, atomic nature of the minerals and the patience, agility, or life-span of the walker. Of course the boundary in question is not static, and the motion of the tides renders attempts to measure the lengths for small λ meaningless, but halting the tides requires no greater a suspension of belief than do many of the models we use out of expediency in science, and it serves our argument well.

A more sedentary way of estimating $L(\lambda)$ is to walk a set of dividers of separation λ along the line on the map. Two steps in this construction are presented in Figs 1.9(b) and (c) and a summary of the dependence of $L(\lambda)$ on λ over the scale range $\lambda = 0.2-L'\ 0.025\ L'$ (where L' is the maximum straight cross-section of the island) is given in Fig. 1.10.

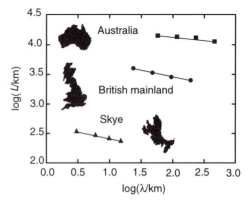

Fig. 1.10 The results of the construction shown in Fig. 1.9 for mainland Britain and the Isle of Skye. We also include the results of a similar analysis for the coastline of Australia, which is relatively smooth at these length-scales.

The empirical law embodied in Fig. 1.10 is known as Richardson's Law after the man who first drew attention to this scaling relationship. This law may be expressed algebraically as follows:

$$L(\lambda) = \beta\ \lambda^{1-\alpha} \qquad (1.12)$$

where α tends to the value 1 for a smooth curve such as a circle, and is greater than 1 for a rough curve; β is an empirical constant for a real object. As the apparent irregularity of the line increases, so α increases. The length may also be expressed as

$$L(\lambda) = n(\lambda)\ \lambda \qquad (1.13)$$

so it appears that

$$n(\lambda) = \beta\ \lambda^{-\alpha} \qquad (1.14)$$

There are similarities between expressions 1.12 and 1.14 and those we derived for the case of the rough surfaces (1.7 and 1.6, respectively). In both cases, α gives some measure of the irregularity of the surface or line. It also appears that as the size of the yardstick or tile approaches zero, so the measure of the length or area approaches infinity. How then may we quantify such an 'area' or 'length'? If we refer back to Fig. 1.1, we see that if we wish to measure the extent of a D-dimensional object, we cover it with n smaller D-dimensional objects and count n. If, instead, we tried to quantify an area using a probe of the *wrong* dimension we get a measure that is either zero or infinity. Thus, if we try to tile an area with lines, or measure a length using tiles, we get a nonsensical answer. We don't usually worry about using a yardstick with the appropriate dimension because it is intuitively obvious. However, for the irregular objects we have considered, it appears that in order to get a finite measure of 'area' or 'length' we must cover them with objects measured in units of length or area raised to the power of $+\alpha$. This is readily appreciated by multiplying out eqn 1.4 with this new form of tile.

$$A(\sigma) = n(\sigma)\, \sigma^{+\alpha} = \beta\, \sigma^{-\alpha}\sigma^{+\alpha} = \beta \qquad (1.15)$$

If we apply the same manipulations to the expressions for $A(\sigma)$ couched in terms of λ, we find that

$$A(\lambda) = k\, n(\lambda)\, (\lambda^{+\alpha})^2 = \beta'\, \lambda^{-2\alpha}\, \lambda^{+2\alpha} = \beta' \qquad (1.16)$$

For the 'length' of the coastline, if we substitute $\lambda^{+\alpha}$ for λ in eqns 1.9 and 1.10 we find that

$$L(\lambda) = n(\lambda)\, \lambda^{+\alpha} = c\, \lambda^{-\alpha}\, \lambda^{+\alpha} = c \qquad (1.17)$$

If we use this definition of the dimension of an object, fractional values may result. Most people find the idea of a fractional dimension unsettling when they first encounter it; many continue to find it unsettling after several encounters. In order to put the reader more at ease at this point, we must look more carefully at what we mean by 'dimension'.

1.2 Three 'dimensions'

The definition of the dimension of an object that we first used was the number of coordinates required to define the position of a point contained by that object; this is called the *Euclidean* dimension and it is given the symbol E. It must take the integral values 1, 2, and 3 for a line, surface, and solid respectively. A second definition of dimension, called the topological dimension, D_T, hinges on the way in which an object must be cut for it to be divided into two parts. A point is indivisible and has dimension 0, while a line requires removal of a point to be severed and is given a dimension 1. The cut required to divide a surface must take the form of a line at least, while division of a solid demands a cut in the form of a surface. The increasingly stringent conditions for division of an object are reflected in increasing integral values of D_T.

The third type of dimension is one that we have loosely called the 'measuring' dimension. We have seen that if we raise our yardstick to the power of the dimension of the object to be measured, we may quantify its size by counting the number of measuring units used. We have also seen that if we use a yardstick raised to the wrong power, we get an answer that is either zero or infinity. We define a third type of dimension as the index to which a linear yardstick must be raised for the measure of the object to be finite; if the value is larger or smaller than this, a nonsensical value will result. This measuring dimension is formally called the Hausdorff–Besicovitch dimension for objects in general, and is given the symbol D_{HB}. More commonly, when D_{HB} takes fractional values which are larger than the topological dimension of the object, we give it the symbol D and call it the *fractal* dimension. Objects for which $D > D_T$ are called *fractals*.

When we are first introduced to the concept of a dimension, it is applied to regular or smooth objects and E, D_T and D_{HB} take the same integral value; determination of D_{HB} implies the same value for E. In the case of an irregular boundary, the different definitions do *not* yield the same value, and we may no longer equate the numerical value of D_{HB} with E or D_T. Let us see how this applies to the rough silica gel surfaces. The surfaces clearly exist in a three-dimensional world with $E = 3$; they have length and breadth and depth. Topologically they are surfaces, albeit very irregular ones: if they were made out of an elastic sheet, they could, in principle, be stretched out to become a smooth surface of infinite area. D_{HB}, however takes the values between 2.9 and 3.0 for various silica gels measured with probe molecules of surface areas between 1.2 and $1.8\,\text{Å}^2$.

In general the following inequality holds:

$$E > D > D_T \tag{1.18}$$

The same principles apply to the length of the British coastline. The image exists on the page as an object with $E = 2$ and $D_T = 1$, while D takes an average value of 1.26 over the length-scale range of at least 25 km to 200 km.

It is important to appreciate that these values of D are limited to a finite range of length-scales for real materials. In the case of the adsorption of molecules by a surface the lower limit is set by the discreet atomic nature of the surface and if the probe molecule gets too large it may become greater than any pore and ultimately larger than the surface itself. Further, it is possible that as the length-scale changes, D changes too. The estimate of D for the British coast may be different when λ ranges from 10^3 to 10^5 m compared with the values obtained when λ is of the order of 10^{-6} m, and it is the texture of the rocks that is being probed. This last estimate will vary from place to place as the nature of the indigenous rocks changes.

1.3 Deterministic fractals and the similarity dimension

The fractals that we have discussed so far are natural objects, and we shall see in Chapter 2 that their growth involves an element of randomness. We may also construct fractals with a regular structure using mathematical formulae and in so doing we reveal more clearly another facet of fractal objects. Such fractals are called *deterministic* fractals because their form is entirely determined by the application of a simple repetitive or iterated process.

Consider the sequence of pictures in Fig. 1.11. Starting with an equilateral triangle, we place a similar triangle of one third the edge on each side. The three straight lines of the perimeter have each been replaced by four smaller lengths whose individual length is one third that of the original. This process is repeated infinitely.

Fig. 1.11 Stages in the creation of a triadic Koch island.

The result is known as the triadic Koch island – 'triadic' because the relative scale of the straight lines between successive iterations is 1/3 and 'Koch' after the mathematician, von Koch, who first described it. The perimeter of the island comprises three lines that are created by taking a straight line, called the *initiator*, replacing it with the motif shown in Fig. 1.12(a), called the *generator,* and repeating this operation as depicted in Fig. 1.12(b)–(e). In the infinite limit, the curve is called the triadic Koch curve. It has several unusual properties.

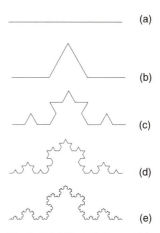

Fig. 1.12 (a) The initiator and (b) generator for the triadic Koch curve plus several successive generations (c) – (e).

(a)   It is infinitely long: $L(\lambda)$ does not tend to a finite limit as λ is reduced.

(b)   It is infinitely spikey: every point on the curve is a corner and there are no smooth parts where one can draw a tangent.

(c)   It has an obvious *dilational* symmetry, that is, if we magnify part of the curve, the enlarged portion looks just like the original. Put another way, as the resolution of the observation is increased, the object continues to look the same. We call such objects *self-similar.*

The natural or *random* fractals with which we started the chapter also have a dilational symmetry over a finite range of length-scales. The scaling analysis which we used to quantify their extent is one manifestation of this, but it has a statistical rather than an exact nature. The photographs of a cauliflower and successively smaller florets in Fig. 1.13 demonstrate this phenomenon: unless we put a linear scale beside it we do not know how large the portion in a photograph is. In geology, a similar phenomenon leads to the inclusion of a familiar object such as a hammer in photographs of specimens. In the case of the deterministic fractals the precise nature of the dilational symmetry provides a means of calculating D quickly.

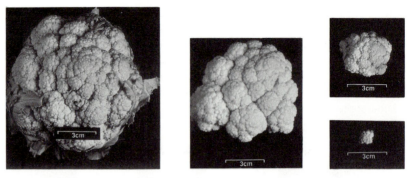

Fig. 1.13 Photographs of a cauliflower taken with increasing magnification.

Conventional E-dimensional objects have a dilational symmetry too. If we halve the edge of the tiles in Fig. 1.1, we find that the number $n(\lambda)$ required to cover the original object increases by 2^E, with $E = 2$. In general, if we reduce the size of the length-scale by a factor R for an E-dimensional object, the number N of new yardsticks, or tiles or building blocks required to 'cover' the original is given by

$$N = R^E \qquad (1.19)$$

In the case of the triadic Koch curve, each step of the process involves replacing a length λ by $N = 4$ lines of length $\lambda/3$. If we rearrange eqn 1.19 and allow E to be non-integral,

$$D = \log N/\log R \qquad (1.20)$$

and substituting for N and $R = 3$, we find $D \approx 1.262$.

The procedure we have described for the generation of the triadic Koch island may be adapted to produce a wide range of curves with $1 < D < 2$. Simple variations are considered in the problems at the end of the chapter. The iterative process may be extended to create a great variety of objects with fractal dimensions ranging from 0 to 3. In the following section we will concentrate on those types of fractals that enlighten our understanding of the structures and chemical processes considered later. Further examples may be found in the book by Mandelbrot to which we refer at the end of the chapter.

Dust and carpets, foams and sponges
Let us first consider the dissection rather than the decoration of a line joining two points. Instead of replacing a straight line with a longer, more complex set of lines, we may remove some of it. Consider the result of excising the middle third of the line, and repeating this. Successive generations, or degenerations if you prefer, are hard to visualize unless we give the line some

substance and represent it with a bar. We show five steps in the formation of this object in Fig. 1.14.

Each step involves replacing the length λ by $N = 2$ lengths $\lambda/3$ so $D = \log2/\log3 \approx 0.6309$. Topologically, the result of an infinite number of iterations is a set of points with $D_T = 0$, and is called a *dust*. This particular type of dust, sprinkled along a line, is called a triadic *Cantor* dust – 'triadic' because of the division into three parts in its construction, and 'Cantor' after the mathematician who invented it. The Cantor dust may have other degrees of excision and hence D. When viewed with a finite resolution it appears to be composed of lines, but when the magnification is increased, any line that appeared to be complete, is found to be divided.

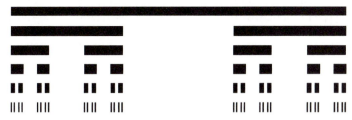

Fig. 1.14 The first five generations of the triadic Cantor dust.

We shall encounter no examples of real materials with this form, but we note in passing that it may be used to represent noise. Suppose we were to monitor the activity on an electrical transmission line when no signal was propagating. The line also transmits spontaneous noise such that a record of activity against time resembles the distribution of mass in the Cantor dust. When monitored with coarse time resolution there will appear to be periods of activity and periods of inactivity. Closer inspection of the active times will reveal shorter inactive regions: as the remaining active periods are scrutinized more closely, so further noise is revealed.

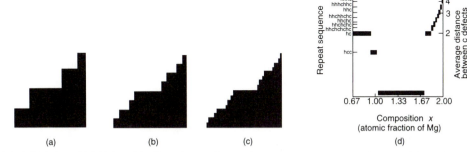

Fig 1.15 (a) 2nd, (b) 3rd, and (c) 4th level Devil's staircase. Successive changes in the repeat distance of the unit cell in the alloys $Mg(Zn_{(1-x)}Ag_x)_2$ or $Mg(Cu_{(1-x)}Ag_x)_2$ are shown in (d) as a function of composition x.

There is an alternative representation of Fig. 1.14 that finds an application in the physical world, and that is formed by plotting the mass of the object against its length. Let us do this for the 2nd, 3rd, and 4th generations in Fig. 1.15 (a) - (c).

In the infinite limit, this shape is known as a Devil's staircase. Its steps are infinite in number and a proportion of these are infinitesimally small in width. It may be used to represent the cross-section of a rough surface. It may also be used to describe certain types of phase transitions in condensed matter. For example, the structure of some intermetallic alloys is finely balanced between hexagonal (h) and cubic (c) closest-packing of atoms. As the composition of the alloy is altered, the electronic structure changes, and with it the balance of forces that determine the packing sequence. However, it is common to observe a gradual change in structure from pure hexagonal (hhhh...) to pure cubic (cccc...) through a series of mixed stacking polytypes in which the repeat distance may be a large multiple of the simple h or c unit cell. If we plot the reciprocal of the repeat distance against the composition we may obtain a graph similar to that in Fig. 1.15(d). This behaviour is common to many crystals, liquid crystals, and magnetic materials in which the structural order arises from a competition between forces with different ranges.

We may decorate or dissect surfaces as well as lines to produce fractal objects. Consider the initiator and generator drawn in Fig. 1.16(a) and (b):

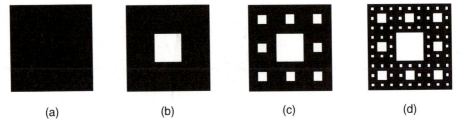

(a)   (b)   (c)   (d)

Fig. 1.16 The initiator and and generator of the Sierpinski carpet with the 2nd and 3rd generations.

We replace the square of side λ by $N = 8$ smaller squares of side $\lambda/3$. D is then $\log 8/\log 3 \approx 1.893$. The product after N iterations is known as the Nth Sierpinski carpet after the Polish mathematician who introduced it. We reproduce the second and third iterations in Fig. 1.16(c) and (d) respectively. A wide range of carpets with $1 < D < 2$ may be produced by removing a square of side $(1 - 2R)$ from the centre of a square of unit side, leaving a ring of $4(1/R - 1)$ squares of side R behind; the fractal dimension is then given by eqn 1.20 as $\log(4(R/3 - 1)/\log(R))$.

Fig. 1.17 Initiator, generator, and two successive stages in the preparation of the Sierpinski gasket.

Sierpinski also lends his name to a second form of dissection of a surface, the Sierpinski gasket, in which an equilateral triangle has a central inverted triangle removed. The three triangles remaining each have half the edge of the original. D is therefore $\log 3/\log 2 \approx 1.585$. We reproduce the initiator, generator, and two successive steps below. Topologically, the Sierpinski gasket and carpet are, in the infinite limit, curves: $D_T = 1$. Their form is reminiscent of the cross-section of porous or cellular solids such as bones and microporous filters – though of course these materials also have a random element. We consider some of these materials in Section 1.5.

The dissected carpet may be extended to higher dimensions to produce curves and surfaces with $1 < D < 3$ if we dissect a solid. Consider the case where the initiator is a cube and the generator is a cube which has been cored by removing a central cube whose side is 1/3 that of the original; $N = 26$ cubes of edge $\lambda/3$ remain, and $D = \log(26)/\log(3) \approx 2.966$ if this process is repeated infinitely. The object that we create contains a distribution of disconnected cubic voids whose volumes range from $(\lambda/3)^3$ downwards, and is one of a family of objects that have been called *foams*. We cannot illustrate this feature of its geometry unless we cut it open and display a cross-section. However, if our generator possesses a void which opens to the outside surface, not only can we observe some of the internal structure, but we may also produce a solid in which the voids are connected to each other. We call such an object a *sponge*. A solid sponge may be constructed by removing the cruciform solid depicted in Fig. 1.18 from the centre of a cube, leaving $N = 20$ cubes of edge $\lambda/3$ behind.

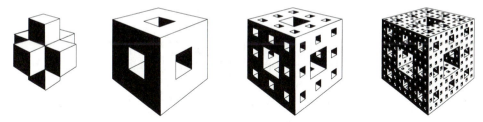

Fig. 1.18 Cruciform core and the first few generations of the Menger sponge.

The object created after an infinite number of iterations is called a Menger sponge, and has dimension $D = \log(20)/\log(3) \approx 2.727$. Topologically, the object is a surface and has no substance: it is all bone and no flesh. It resembles some of the porous solids with which we started the chapter. The reader might also consider that it resembles the zeolite too (Fig. 1.3) in that it has a high symmetry. However, it differs from any zeolite in that it possesses a broad distribution of pore sizes. In this respect it is more like a lump of Emmental cheese, though it differs from that too because the Swiss cheese has a lower limit to the hole size while the Menger sponge has an infinite range. Note that the faces of the Menger sponge are triadic Sierpinski carpets, and that a line drawn along a diagonal of a face represents a triadic Cantor dust (Fig. 1.18).

1.4 The box-counting dimension

The Koch curve is, in some respects, a model for our rough coast. It has a similar fractal dimension to the coast of mainland Britain, and we see under increasing magnification smaller and smaller 'bays'. We should also be able to quantify its length with the walking technique. The results of such an analysis are presented in Fig. 1.19.

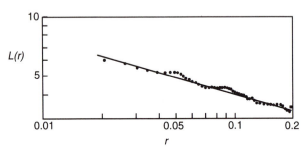

Fig. 1.19 Determination of the fractal dimension of the perimeter of the triadic Koch island of Fig. 1.14 through the walking technique.

Fig. 1.20 Enlargement of part of coastline of Fig. 1.9, demonstrating that the length of the walk depends on whether the walker cuts across an inlet at the first opportunity (arc $a - b$) or keeps to the same shore for as long as possible (arc $a - c$).

We see that the symmetry of the object produces a periodic fluctuation in the plot of $L(\lambda)$ against λ. Furthermore, in both this case and that of the coast, there is some ambiguity in the way in which the walk is performed.

Consider Fig. 1.20, which is an enlargement of part of the coastline of Fig. 1.9. If we walk the dividers into an inlet whose mouth is comparable in width to λ and which is longer than λ, we may have a choice of 'next' points in our walk. Suppose the walker stands at point a in Fig. 1.20 and the two points on a radius λ from a are b and c. If the walker is afraid of getting her feet wet and hugs the coast, she would reach c first, and explore the inlet. This is the 'out-swing' construction, in which the dividers are always swung from the inside of the object. If she preferred to stride across the inlet, she would cut across to b. The latter 'in-swing' construction means that she will never explore an inlet unless λ is smaller than the width of its mouth. The out-swing construction produces a higher value of D.

There is an alternative method of evaluating D which is free from these problems. Suppose we place a grid of square boxes of side λ over the boundary to be quantified and mark those cells that contain part of it. The string of n marked cells form a ribbon of area $n \lambda^2$ so the length of the boundary measured at this resolution is $n \lambda$. As we reduce λ we expect $(n \lambda)$ to increase, and to scale in the same way that $L(\lambda)$ scaled with λ in Fig. 1.10. The dimension D calculated with this box-counting technique is:

$$D = \log n/\log(1/\lambda) \qquad (1.20)$$

We show the first and a few successive stages in this procedure for the coastline of the mainland of Great Britain in Fig. 1.21. A plot of the results is shown in Fig. 1.22, with the lower cut-off for λ being the width of the line. If λ falls below this limit the method senses that the line has a finite width and starts to treat it as a surface.

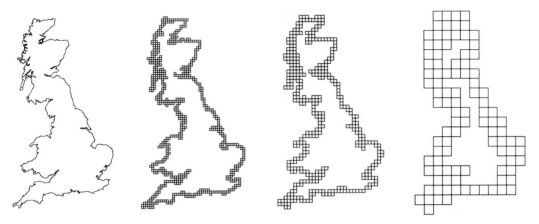

Fig. 1.21 Several stages in the calculation of *D* for the coastline of the mainland of Great Britain using the box-counting method.

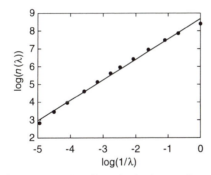

Fig. 1.22 Calculation of the box-counting dimension for the coastline of the mainland of Great Britain.

If we apply this method to the microscopic structure of real solids, the lower cut-off for λ is the radius r_0 of the molecular or atomic building blocks. For a conventional *E*-dimensional object, the number $n(r)$ of particles within a radius r in the molecular fragments taking the form of a linear chain, disc, and sphere displayed in Fig. 1.23 is given by

$$n = p \, (r/r_0)^E \tag{1.21}$$

where p is the number density or number of building blocks per unit length, area, or volume for $E = 1$, 2, and 3 respectively. It is determined by the way in which the molecules pack within the object. For the three simple shapes in Fig. 1.23(a)–(c), p is equal to 1, $\pi/2\sqrt{3}$, and $\pi/3\sqrt{2}$, respectively. The branching polymeric structure of Fig. 1.23(d) may also be treated in this way with E replaced by D for large n; p still depends on the way in which the monomers pack within the strands of the material. The density of the solid as a function of its radius is assigned the symbol $\rho(r)$ and is given by:

$$\rho(r) \approx n(r)/r^E = p \, r^{D-E} r_0^{-D} \tag{1.22}$$

Thus, if $D < E$, $\rho(r)$ will decrease with r. Such an object is known as a *mass fractal*. Although the density of the object no longer behaves as an intensive property the substance from which it is made has a uniform density. $\rho(r)$ *would* remain an intensive property of the fractal if its volume was also calculated using a similar procedure to the way in which the mass is calculated, rather than also counting all the space between strands of polymer as part of the overall volume. In practice, when we measure the mass or matter distribution in an object, we use methods which naturally probe the fractal scaling properties, as we shall see in Chapter 3.

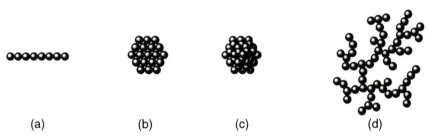

| (a) | (b) | (c) | (d) |

Fig. 1.23 Packing of spheres to produce (a) a line, (b) a disc, (c) a sphere, and (d) an irregular branched object embedded in $E = 2$ dimensions.

The form of a mass fractal may be related to the network of pores in a highly porous solid. Consider the mass fractal at the top of Fig. 1.24, and its doppel-gänger at the bottom, which could represent the distribution of air in a porous solid. We call such a material a *pore fractal* and note that the mass of any substance pumped into the pores scales with the radius r of the host material as r^D.

1.5 The relationship between perimeter, area, and volume for fractals

Commonly, an object with a perimeter L and area A shows the scaling relation

$$L/A^{1/2} = k \tag{1.23}$$

Fig. 1.24 Mass fractal and pore fractal for the embedding dimension $E=2$.

Similarly, A and volume V may be related by

$$A^{1/2}/V^{1/3} = k' \tag{1.24}$$

where k' depends on the shape of the object and is independent of the yardstick length λ. You may wish to verify this relationship by considering the precise form of the area of a disc or the volume of a sphere. For a fractal curve enclosing an area, this ratio will only be independent of λ if $L(\lambda)$ is raised to the power of $1/D$ rather than $1/E$ i.e.

$$(L(\lambda))^{1/D}/A^{1/2} = k' \qquad (1.25)$$

where k' is now independent of λ but *does* depend on the shape of the coast. Thus, the length of the coast estimated with the box-counting technique diverges as λ is reduced while the area tends to a finite limit. Similar relations hold for other fractal objects: the drainage basin of a river has a finite area, but the length of the river within that basin may show fractal scaling over a wide range of values of λ. The volume of the brain of a large mammal scales with linear dimension in a conventional manner, but the surface area of complex, folded structure when stretched out obeys the scaling relation

$$V^{1/3} \sim A^{1/D} \qquad (1.26)$$

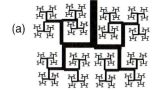

(a)

with D between 2.73 and 2.79. Other biological boundaries show similar relationships. The mass of an organism and its demands on resources – oxygen, water, and fuel – scale approximately as the cube of one of its linear dimensions. The intake of nutrients, respiration, and cooling through evaporation require exchange of matter through biological membranes, and are proportional to their surface area. Consequently, if the organism and its organs obey Euclidean scaling laws, the degree to which its demands on resources may be met falls off as the reciprocal of the linear size. Clearly, therefore, there is a need to greatly enlarge the extent of active boundaries~as the size of an animal increases; lungs and blood vessels adopt intricate surface structures such that the ratio of surface area to mass falls off as a smaller power of r. The deterministic fractal in Fig. 1.25(a) is suggestive of the structure of a lung's bronchial system viewed in cross-section (Fig 1.25(b)). The tubes in kidneys and hearts show similar properties.

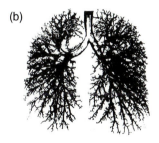

(b)

The strength of a bone scales with its cross-sectional area while the mass scales with the volume so that as an animal's size increases, the ability of its bones to support its weight falls off as the reciprocal of its linear dimension, unless the bones occupy a greater proportion of the animal's bulk, or the ratio of cross-sectional area to mass scales as a lower power of r. The latter condition is met by constructing the bones out of a very porous material.

Fig. 1.25 Biological fractals: (a) is a deterministic fractal that is reminiscent of the structure of a lung (b).

The porous solids that are used in catalysis are often in the form of powders whose surface obeys fractal scaling laws over several orders of magnitude of the probe molecule. The mass of samples of these powders depends on the volume in a conventional manner.

$$M = \rho V = \rho r^{E} \qquad (1.27)$$

with $E = 3$. Thus, the ratio of surface area to mass is given by

$$A/M = r^{D-3} \qquad (1.28)$$

In practice, many objects that have been grown with fractal structures appear to have $D = 3$ when analysed in this manner. Their tenuous structure renders them weak when subjected to the mechanical stress of grinding, breaking them into more compact particles which are relatively smooth. A material with a rough surface or in the form of a powder for which the surface area A is proportional to r^D, where r is the resolution with which it is studied, is known as a *surface fractal*.

1.6 Summary

Fractal objects have a dilational symmetry which is exact for a deterministic fractal and of a statistical nature for the random fractals found in nature. For a deterministic fractal, the number of objects of linear extent r required to cover the fractal at each step of its creation is equal to r^{-D} where D is a fractal dimension which in this case may also be called the similarity dimension. D may be distinguished from the topological dimension D_T and the Euclidean dimension E, because it may take non-integer values (ranging between 0 and 3) which conform to the inequality $E \geq D \geq D_T$.

The number of objects of linear extent r required to cover a natural fractal scales as r^{-D} over a finite range of r. The object may be a rough surface, or an irregular cluster or a porous solid for which the extent of the surface, or mass, or pore volume scales with the radius of the object as r We call such objects mass fractals, surface fractals and pore fractals respectively. The area per unit mass of a fine powder which is a surface fractal scales with the radius as r^{D-3}.

In Chapter 2 I shall describe how fractals may be grown, in Chapter 3 I shall outline methods of characterization and, finally, in Chapter 4, I describe how the fractal properties of materials influence some of their chemical behaviour.

Further reading

Thompson, D'A. W. (1992). *On growth and form.* Cambridge University Press.

Mandelbrot, B.B. (1983). *The fractal geometry of nature.* W.H. Freeman, New York.

Avnir, D. (ed.) (1990). *The fractal approach to heterogeneous chemistry.* John Wiley and Sons, Chichester.

Kaye, B.H. (1989). *A random walk through fractal dimensions.* VCH, Weinheim.

Feder, J. (1988). *Fractals.* Plenum, New York.

Iler, R.K. (1979). *The chemistry of silica.* John Wiley and Sons, New York.

Bak, P. (1986). The Devil's Staircase. *Physics Today,* December 39.

West, B.J. and Goldberger, A.L. (1987). Physiology in fractal dimensions. *American Scientist,* **75** 354.

Many of the fractal forms described in this chapter may be generated and displayed using simple computer programs. You may wish to write these yourself, or to extract software from one of the archives linked to the international electronic network INTERNET which may be accessed through main-frame computers at universities or through a modem at home. There are many books that contain software and some listing of computer code for the generation of fractal images. These include the following.

Wegner, T. and Peterson, M. (1991). *Fractal creations*. Waite Group, Mill Valley, California.
Barnsley, M.F. (1992). *The desktop fractal design system*. Academic Press, San Diego.
Oliver, D. (1992). *Fractal vision*. Sams, Carmell, Indiana.

Problems

1.1. The length of the coastline of the British mainland was paced out by a lone walker as 10^4 km. Use Fig. 1.10 to estimate what is the step length that corresponds to this perimeter, and comment on the result.

1.2. Calculate the fractal dimension for the perimeter of the Koch island whose initiator is a square and whose generator is depicted in Fig. 1.26 with (i) $\theta = 90°$, (ii) $\theta = 120°$ and, (iii) $60°$. Note that the three lines separated by dots in this figure have equal lengths. You may find it helpful to note the cosine rule, which states that for a triangle whose edges have lengths a, b, and c and for which the angle between edges a and b is γ, $c^2 = a^2 + b^2 - 2a.b \cos\gamma$.

1.3. Draw a generator for a Koch island whose initiator is a square and whose coastline has a fractal dimension of 3/2.

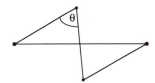

Fig. 1.26 Generator for the Koch island of problem 1.1.

1.4. The number of moles n for monolayer coverage of a porous form of charcoal called carbon black with nitrogen and a series of aromatic hydrocarbons of different cross-sectional area σ is given in Table 1.1. Calculate the fractal dimension of this surface over this range of length-scales and estimate the effective area of naphthalene given that 0.80 mmol is required for monolayer coverage per gram of carbon black.

Table 1.1 Number of moles n of nitrogen or selected aromatic hydrocarbons of different area σ for monolayer coverage of carbon black.

Adsorbate	σ (Å^2)	n (mmol g^{-1})
Nitrogen	16.2	3.33
Benzene	35.2	1.30
Anthracene	70.7	0.65
Phenanthrene	68.8	0.65

1.5. The activity per unit mass of a powdered catalyst depends on the particle radius r in the manner given by eqn 1.28. Measurements performed on different samples with graded particle size suggest a range of values for the exponent D that commonly range from slightly larger than 2.0 to below 1. The values greater than or equal to 2.0 are compatible with the idea that activity is simply proportional to area, which may scale as a flat or fractally rough object. How might exponents significantly *smaller* than 2.0 be rationalised? Hint: the activity of vertices or edges of the particles may be far more significant than that of the flat regions. How will the number of sites contained in such features scale with r?

2 Fractal growth

2.1 Occurrence of fractal objects in nature

It is clear from the preceding chapter that objects which obey fractal scaling laws over a range of length-scales are common in nature, and that their origins can be very diverse. The weathering or erosion of the interface between a rock and the elements –heat, wind, water, and grit –gives rise to fractal coasts and to the drainage basins of rivers, to karst scenery and the labyrinths of underground cave systems worn in porous rocks. At shorter length-scales we have seen that fractal objects are formed in a variety of aggregation processes such as the formation of silica gel. Finally, we noted the incidence of fractal structures in organisms, the network of tubes in lungs and hearts and livers which maximize the surface area to volume ratio in animals, and the formation of porous load-bearing structures in the skeletons of flora and fauna which maximize the ratio of the cross-sectional area to mass. The comparison made in Fig. 2.1 between fractal objects created by electrochemical aggregation and erosion provides a visually appealing, though highly unrigorous argument for a common basis for their growth.

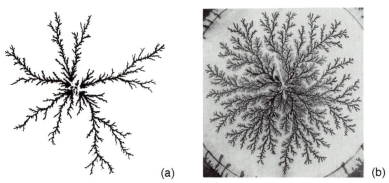

(a)
(b)

Fig. 2.1 (a) Zinc grown by electrodeposition and (b) the pattern formed when water is injected into the centre of a thin layer of clay held between two glass plates.

In this chapter we examine some of the conditions under which fractal growth may occur and see that there are common features in many, though not necessarily all, of the growth processes. Throughout the chapter we describe simple experiments which the reader could perform in a modestly equipped laboratory. We begin by considering a simple model for the growth of a solid under conditions that lead to either dense or fractal products.

2.2 Electrochemical growth of fractals

One of the easiest methods of creating a fractal object is the electrochemical deposition of a solid. In Fig. 2.2(a) we show a simple cell for the deposition of metallic silver. The electrodes are two strips of silver paint of the sort used for good thermal or electrical contact by electronic engineers. The electrolyte is made up from 20 ml of distilled water with three drops of ammonium hydroxide, and a drop is added between the silver strips. A microscope cover slip is placed on top and excess water wiped away. The cell is sealed with an epoxy resin and left to dry. Electrolysis occurs when a voltage is applied across the terminals: a 4.5 V battery will usually suffice, although some experimenting with voltage and strip separation and electrolyte composition may be necessary. Growth of fern-like silver deposits at the cathode may be observed in a few seconds with a microscope; alternatively, the slide may be cut so that it fits in a projector and growth may be observed on a screen.

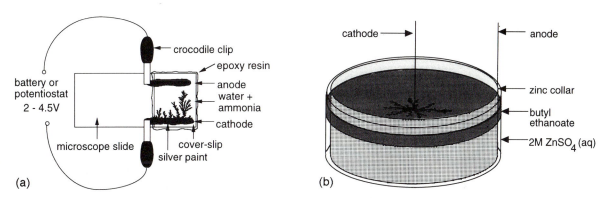

Fig. 2.2 Simple electrochemical cells for the observation of fractal growth of (a) silver and (b) zinc.

Electrochemical growth of fractal metal patterns embedded in $E = 2$ dimensions may also be conducted with the apparatus portrayed in Fig. 2.2(b). An aqueous solution of 2M $ZnSO_4$ sits in a glass basin of about 20 cm diameter and 10 cm depth, and is covered with a few millimetres of butyl ethanoate. A sheet of zinc metal is cut and bent into a ring which lines the inside of the basin, and this provides the anode; a cathode is supplied in the form of a copper wire or graphite pencil 'lead' whose diameter is about 0.5 mm. A potential of about 5 V is put across the electrodes using a potentiostat, and the cathode is lowered gently into the centre of the liquid with the aid of a laboratory jack and a retort stand. As the cathode comes into contact with the aqueous solution, growth of zinc metal starts, and is confined to the plane of the interface between the two fluids until the pull of gravity draws it into the aqueous layer. The evolution of the pattern may be followed by conducting this experiment on the top of an overhead projector, and it typically takes five or ten minutes to grow a pattern of diameter 10 cm, though you should experiment with the voltage and concentration of zinc in the solution because the growth will also depend on the ambient temperature.

A more sophisticated and informative experimental set-up for the electrochemical growth of a fractal object is based on the oxidation of pyrrole

(C_4H_4N) at an anode in a thin cell. The cathode of the cell is provided by a 10 μm layer of gold which is evaporated onto a glass plate leaving a 2.5 cm diameter hole in the centre. A second glass plate rests on this, and the anode is a fine gold wire pushed through a hole in the centre and cut flush with the underside using a razor blade. The cell is completed by filling the thin cavity with an electrolyte which is a solution that is 0.1 M in both pyrrole and AgO toluene sulphonate in acetonitrile. Electrochemical polymerization of pyrrole occurs at the anode when a variable voltage source is connected across the electrodes.

Polymerization may occur above a threshold voltage of about 0.8 V; it follows the removal of two electrons per molecule of pyrrole and is accompanied by the loss of two hydrogen ions. The polymer is oxidized further to the tune of one electron lost for approximately three monomer units. The anode process may be summarized by the following reaction scheme:

$$3n \begin{array}{c}\text{(pyrrole)}\end{array} \longrightarrow \left[\text{(polypyrrole)} \right]_n + 6ne^- + 6H^+$$
$$\left[\text{(polypyrrole)} \right]_n^+ + 7ne^- + 6H^+ \qquad (2.1)$$

Reduction of silver ions to silver occurs at the cathode.

At low voltages the polymer grows as a dense object whose mass scales with its radius r as r^2. As the voltage is increased above 1.0 V, the growth becomes less compact, adopting a tree-like pattern at intermediate voltages, with some degree of regularity in the angles that smaller branches make with larger ones.

(a)

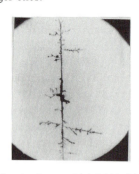
(b)

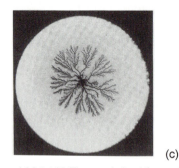
(c)

Fig. 2.3 Polypyrrole grown electrochemically at voltages of (a) 0.9 V, (b) 1.5 V, and (c) 6.0 V. The anode is at the centre of the growth and is not visible; the cathode is the dark ring that frames the picture.

Above about 3.0 V, even this regular character is lost and random branching occurs to produce a solid that appears to be a fractal. There is no observable change in the form of the growth for an applied voltage above about 6.0 V. We reproduce a series of polymers grown at different voltages in Fig. 2.3. Once grown, the product may be photographed and the image analysed using the box-counting technique. For voltages that produced fractal growth, the mass is found to scale with the radius raised to the power of 1.74±0.03 for length-scales ranging from 0.3 to 6.0 mm.

What is the reason for the change in the growth pattern? To answer this we must look more closely at the way in which the polymer grows. Pyrrole molecules diffuse through the solution until they collide with the surface of the polymer or the anode. We represent the anode reaction as

$$R = O + ne^- \qquad (2.2)$$

where R and O are the reduced and oxidized forms of pyrrole. The rate for the forward (f) and reverse (r) reactions, k_f and k_r, may be expressed using activated complex theory as

$$k_f = A_f \exp(-\Delta G_f^\ddagger / RT) \qquad (2.3)$$

$$k_r = A_r \exp(-\Delta G_r^\ddagger / RT) \qquad (2.4)$$

where A_i is the pre-exponential factor for the processes $i = $ f or r and is related to the frequency with which the reactants attempt to cross the activation barrier, $\Delta G_i^\ddagger$. Activated complex theory tells us that we can control the value of $\Delta G_f^\ddagger$ and $\Delta G_r^\ddagger$ by changing the free energy of R and O respectively. The sensitivity of $\Delta G_i^\ddagger$ towards such changes depends on the degree to which the activated complex resembles R or O. We express this in terms of the position of the activated complex along the reaction coordinate which in this case may be interpreted as the degree to which the electron is passed between the anode and the pyrrole molecule. The position along the reaction coordinate is quantified with the constant β which is called the transfer coefficient and takes values from 0 to 1, depending on whether the activated complex is like the reactants or products in the forward direction. In Fig. 2.4 we show the development of the free energy of the reaction for β ≈ 0.5.

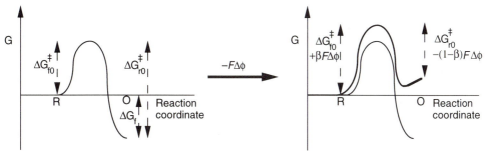

Fig. 2.4 The dependence of free-energy of a redox reaction as a function of reaction coordinate when the potential is changed by Δφ. The symbols are defined in the text.

We may change the free energy of the electron at the anode by changing the potential of the anode relative to the solution by Δφ. This produces a free energy change of $-F\Delta\phi$ per mole of electrons and consequently a change in $\Delta G_r^\ddagger$ of $-\beta \, F\Delta\phi$ per mole. The rate constant k_f then changes to:

$$k_f = A_{f0} \{\exp(-\Delta G_{f0}^\ddagger / RT) + \exp(\beta \, F\Delta\phi / RT)\}$$

$$= k_{f0} \exp(\beta \, F\Delta\phi/RT) \tag{2.5}$$

where $\Delta G_{f0}^{\ddagger}$ is the free energy of activation with no applied potential, and

$$k_{f0} = A_{f0} \exp(-\Delta G_{f0}^{\ddagger}/RT) \tag{2.6}$$

$\Delta G_r^{\ddagger}$ changes by $(1-\beta)F\Delta\phi$ and the rate equation changes accordingly:

$$k_r = k_{r0} \exp(-(1-\beta) \, F\Delta\phi/RT) \tag{2.7}$$

At equilibrium, $[R]k_f = [O]k_r$ and $\Delta\phi = \Delta\phi_{eq}$ so that

$$[R] \, k_{f0} \exp(\beta \, F\Delta\phi_{eq}/RT) = [O] \, k_{r0} \exp(-(1-\beta) \, F\Delta\phi_{eq}/RT) = j_0/F \tag{2.8}$$

where j_0 is the current density in either direction at equilibrium and F is Faraday's constant. If we define the difference between $\Delta\phi$ and $\Delta\phi_{eq}$ as the overpotential η, we find

$$k_f = j_0/F \exp(\beta \, F\eta/RT) \tag{2.9}$$

$$k_r = j_0/F \exp(-(1-\beta) \, F\eta/RT) \tag{2.10}$$

Therefore, if η is large and positive, $k_f \gg k_r$ and $[O] \gg [R]$. At very high voltages, it is almost certain that a pyrrole molecule will be oxidized and stick to the first site it hits. This is called *diffusion limited aggregation* (DLA), and is controlled by the way in which the particle approaches the perimeter of the growing cluster. As the voltage is reduced, there is an increase in the probability that the pyrrole molecules may collide several times before oxidation occurs. Both the free energy of deposition at a particular site and the diffusion process may then influence growth, and this is called *diffusion controlled aggregation* (DCA). Finally, growth may occur when the molecules have had an opportunity to sample so many sites that growth occurs at those which are most favoured by a low free-energy for deposition. This is termed *reaction limited aggregation* (RLA). In the present case, all three types of growth occur, depending on voltage and electrolyte concentration. Fractal growth occurs when there is DLA.

DLA and DCA are widespread phenomena, particularly in the growth of colloids in both liquid or gaseous media. Similar processes govern the growth of irregular solids from saturated solutions and melts. We will consider these and other processes that give rise to fractal growth in the remainder of this chapter. First we consider a simple model for DLA which provides some insight into the nature of fractal growth.

2.3 Computer simulations of DLA

As the speed and memory capacity of computers have increased over recent years, together with ease of access and familiarity, so computer simulations have become increasingly important to scientists. Although such simulations are limited to relatively small collections of particles – perhaps 10^4 to 10^5 for the simplest calculations, and much fewer for complex calculations – they may provide insight where an analytic solution is elusive. Some of the most notable successes have been in the field of fractal growth where a few simple models reproduce most of the features observed in real materials.

In the simplest model, named after Witten and Sander, the scientists who devised it, the region of space in which a cluster grows is represented by a square grid with axes x and y. A tile at the centre of the grid is labelled to represent occupation by the seed of the aggregate to be grown. When DLA occurs in a real material, particles execute random walks, buffeted by other particles in the fluid environment, until some of them strike the growing cluster and bind. Most of these trajectories do not come anywhere near the cluster, and, to save computer time, we ignore them. In the simulation, particles are introduced to the system at a radius which is slightly larger than that of the growing cluster, R_{max}. This is shown as the circle S_1 in Fig. 2.5(a).

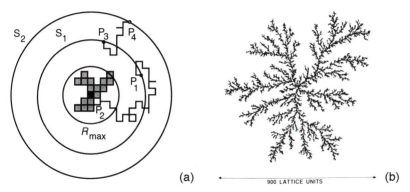

(a)   900 LATTICE UNITS   (b)

Fig. 2.5 (a) Illustration of the growth of a diffusion-limited aggregate on a square lattice showing the outer cut-off limit S_1 and the ring S_2 on which new particles are introduced. We show random trajectories for a particle that does contribute to the growing cluster, and a particle that wanders beyond S_1. In (b) we show the result of such a growth for 5.10^4 particles on a hexagonal lattice.

The particle is then allowed to take a random walk on the square lattice. It does this by means of the computer's random number generator which produces a sequence of random numbers between 0 and 1. If the number lies between 0 and 0.25 the particle is moved in the $+x$ direction; numbers in the remaining three quartiles direct the particle in the other three nearest-neighbour directions, $-x, +y,$ and $-y$. If this particle wanders too far away, beyond a circle S_2 whose radius is arbitrarily chosen to be some multiple of R_{max}, it is discarded, and another particle is introduced at another random position on S_1. If the particle collides with the cluster it is added to it and another particle introduced at S_1. The length of the simulation is limited by the amount of time available on the computer and simulations. Typical cluster sizes range between 10^4 and 10^7 particles. We illustrate the growth process in Fig. 2.5(a) and beside it (b) show the result of such a simulation in which a cluster of 5.10^4 atoms has been grown on a hexagonal lattice.

The fractal dimension may be calculated by determining the manner in which the number of particles at a given radius from the seed scale with radius. It is found to be independent of the symmetry of the grid used: square and hexagonal lattices give the same results within statistical error. However, it is strongly dependent on the dimension of the lattice on which the cluster grows. These results are summarized in the first rows of Table 2.1.

Table 2.1 Fractal dimension of objects grown by DLA and RLA processes, through computer simulations and in real materials. We show the results for aggregation in which only the elementary particles are free to move, as well as the case where the growing clusters may move.

Embedding dimension	Growth method	Particle-cluster aggregation	Cluster-cluster aggregation
3	Computer DLA	2.50	1.80
2	Computer RLA	2.0	1.61
3	Computer RLA	3.0	2.09
2	Electrochemical (Zn)	1.70	
3	Electrochemical (Cu)	2.43	
2	Polystyrene microsphere colloids		1.42
3	Vapour aggregation of Fe		1.70 – 1.90
3	Colloidal aggregation of Au		1.75 – 2.10

We have assumed that the mean-free path of the particles that are aggregating is of the order of, or significantly smaller than, the growing cluster. If the density of these particles is very low, the random collisions that influence the random walk may be very infrequent. The trajectories of the particles will still have random orientations but they will now appear to be straight at length-scales comparable to the cluster radius. This type of growth is known as ballistic aggregation and leads to compact growth of clusters with the same dimension as the embedding dimension – that is, the dimension of the space in which the growth occurs. At some point, these compact clusters may become large relative to the mean-free path of the aggregating particles, and a cross-over to fractal growth is observed.

2.4 Experimental observation of DLA in colloid formation

There are many tangible examples of DLA besides electrochemical deposition. Perhaps the most direct realization of the simple DLA model described above occurs in colloid formation. Typically, small particles suspended in a gas, liquid, or solid diffuse together and bind through a variety of forces. Colloids in which the particles are suspended in a liquid or a gas are called a sol or an aerosol, respectively.

One of the first processes interpreted in this way was the formation of fractal clusters in an aerosol of iron or zinc. In the case of iron, a vapour was produced by passing an intense pulse of electrical current through an iron-plated tungsten filament in a vessel filled with helium gas. The gas restrained the diffusion of the iron atoms and facilitated the formation of small crystallites of radius 3.5 ± 15 nm. These then aggregated to form stringy structures which could be captured on the fine mesh of an electron microscope sample holder and imaged with transmission electron micrography. D was found to be between 1.7 and 1.9 for length-scales that varied by a factor of up to 40.

A particularly good example of colloid growth by DLA in solution is the growth of gold clusters. A sol composed of very regular spheres of gold of 14.5 nm radius is prepared by reducing $Na(AuCl_4)$ with sodium citrate (Fig. 2.6(a)). The surface of the spheres adsorbs citrate anions and stabilises them against aggregation through electrostatic repulsion. If pyridine is now added to the solution, it may rapidly strip away the anions, making the sol particles neutral and initiating clustering. Solids grown in relatively high concentrations of pyridine $(10^{-2} M)$ have a fractal dimension of approximately 1.75 as determined by image analysis of photographs taken by transmission electron microscopy. If much more dilute pyridine solutions are used $(10^{-5} M)$ the citrate anions are only partially stripped away and the probability that particles bind when they first come together is reduced to the extent that RLA prevails and the fractal dimension takes values up to 2.10 ± 0.10.

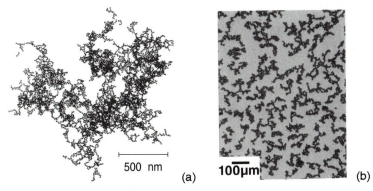

500 nm   (a)   100μm   (b)

Fig. 2.6 Examples of objects grown by DLA: (a) a gold sol and (b) polystyrene microspheres in water.

There are many ways in which we may observe DLA in the laboratory. One simple experiment uses polystyrene spheres with a uniform diameter of the order of $1 - 5 \mu m$ and constrains their motion to a plane by dispersing them in a layer of water held between two thin glass plates which are kept apart by a number of polystyrene spheres of slightly greater diameter. The collection of particles is first stabilized against aggregation with a surfactant: if a small amount of sodium dodecyl sulphate is added, the organic end of the molecule is attracted to the polystyrene so that the ionic end points outwards and provides a repulsive electrostatic force. Aggregation is then triggered by increasing the ionic strength of the solution, allowing the spheres to approach sufficiently close for the short-range attractive Van der Waals potential to produce a bond. When growth is slow, aggregation produces dense, crystalline forms, while faster growth produces clusters with fractal dimensions significantly lower than 3 as illustrated in Fig. 2.6(b). Similar growth may be observed with photocopier toner in place of the polystyrene.

In Chapter 1 we described the results of DLA of SiO_2 in acid and base solutions: primary particles in the form of spheres of SiO_2 undergo DLA when the solution is acidic and form a network that entraps solvent, and this may be dried to produce a xerogel. However, unless special precautions are taken, the drying process usually involves the formation of solvent vapour in the cavities and this coexists with solvent. The surface tension at the interface acts to minimize its area and provides a destruction force that causes the silicate network to collapse. If the solvated gel is heated under pressure to the critical point of the solvent we may force all the solvent to be converted to vapour at once. If the solvent vapour is now pumped away the silicate framework will remain intact, producing a solid of very low density called an aerogel.

Low-density compounds of silica may also be grown in a vapour in a manner similar to that described for the metal aerosols. The most common way of preparing a silica aerosol is to burn $SiCl_4$ with H_2 and O_2 in the presence of an inert carrier gas such as Ar. Spheres of SiO_2 with a well-defined distribution of diameters that lie between 10 and 20 nm are produced in the flame and then aggregate to produce light solids known as fumed silicas, or by trade names such as Aerosil or Cab–O–Sil. Such materials provide important sources of pure and doped silica for optical fibres and are common additives as thickeners in foods and toothpaste.

Our final common example of objects grown by DLA is the soot that forms when hydrocarbons are burned in insufficient oxygen for all of the carbon to be converted to CO or CO_2. The flame converts the hydrocarbon into particles of diameter $10-20$ nm and these may aggregate and fuse to produce clusters similar in appearance to the other DLA clusters in Figs. 1.4, 1.5, and 2.6. The general name for this form of soot is 'carbon black' and it has been used for many years as an industrial adsorbent on account of the high porosity, as well as a versatile catalyst. More recently, there have been concerns about the environmental damage that may be caused by soot in car exhaust emissions, and in particular in the exhaust from an imperfectly–tuned diesel engine. Such materials are suspected to be particularly efficient at adsorbing carcinogens and transporting them into the body in much the same way that respirable silicates such as asbestos compound the carcinogenic activity of cigarette smoke.

The fractal dimension predicted by the computer simulation of DLA is commonly significantly smaller than the values for most real materials grown in this way. The major discrepancy appears to arise from the assumption that growth is centred on a single seed in the computer simulations. Real DLAs generally nucleate at many sites to form small clusters which may grow either through the addition of particles or collision with other growing clusters. The modification of the model to take such processes into account greatly increases the computational time, and has been restricted to growth of relatively small clusters. The model now allows both the primary particles and the clusters to move randomly. Typical values for DLAs grown in this manner for embedding dimensions of 2 and 3 are given in Table 2.1 together with some experimental results.

The computer simulations mimic natural aggregation processes with remarkable success, but what insight do they give us into why fractal objects grow the way they do? Why does DLA favour the growth of protuberances at the expense of depressions? What relation do the colloidal aggregates have to some of the fractal objects in Fig. 2.1.

An intuitive answer to all of these questions lies in Fig. 2.7 which depicts the development of a bump on a flat surface as it is exposed to a flux of randomly walking particles that stick to the surface when they collide. Even if the surface is originally smooth, the random sequence of events that control its development will produce some unevenness. When rough features form, they influence further growth because they make certain trajectories less probable than others. Sites at the bottom of a pit are harder to reach than those at a peak because the wandering particles are likely to be trapped on the walls of the pit. The peaks that grow are themselves subject to the growth of protuberances, so that the spikes that emanate from the seed wander and split to produce ever finer features and statistical dilational symmetry.

The collective behaviour of a set of particles executing random walks controls many of the dynamic properties of fluids. In addition to DLA, the flow of heat and the motion of solute particles in a concentration gradient, as well as the viscosity, depend on such motion. We shall next consider how these transport properties arise from the statistical average of a collection of particles performing random walks.

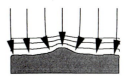
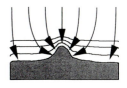
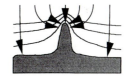

Fig. 2.7 Amplification of irregularities in a surface grown by DLA.

2.5 Random walks and diffusion

Consider a particle of mass m suspended in a bath of fluid at a temperature T. Regardless of m, the particle will have a thermal energy $kT/2$ associated with motion along each of the orthogonal axes along which it is free to move. If we consider just one direction, x, for simplicity, then we may also express the kinetic energy E for motion along x as

$$E = mv_x^2 / 2 \qquad (2.11)$$

so the mean square value of the velocity v_x, denoted $< v_x^2 >$, is given by

$$< v_x^2 >= kT / m \qquad (2.12)$$

If we assume that the particle hops a mean distance δx in a time τ, then

$$\delta x = \pm v_x \tau \qquad (2.13)$$

We will assume that δx and τ are constant for a particular particle in a particular fluid, and that each step is independent of the nature of previous steps. For a collection of N particles, the position of the ith particle after n steps is denoted by $x_i(n)$, which may be expressed in terms of the position of the previous step, $x^i(n-1)$, as follows:

$$x_i(n) = x_i(n-1) \pm \delta x \qquad (2.14)$$

The mean displacement after n steps requires summation of $x_i(n)$ over the N particles and division by N:

$$<x(n)> = 1/N \sum_{i=1}^{N} \{x_i(n-1) \pm \delta x\} \qquad (2.15)$$

In Fig. 2.8 we plot an example of $<x(n)>$ as a function of n generated either with a die or the random number generator of a pocket calculator. The random walk may make relatively large excursions, but will always pass back through zero. We note in passing that the distribution of points along the N axis at which the walk crosses through zero displacement is distributed as a Cantor dust (Section 1.3). The term δx in eqn 2.15 sums to zero because the positive and negative signs are distributed randomly so

$$<x(n)> = 1/N \sum_{i=1}^{N} x_i(n-1) = <x(n-1)> \qquad (2.16)$$

The average displacement remains fixed at the starting position. The spread of particles is given by the mean of the square of the displacement, in turn given by

$$x_i^2(n) = (x_i(n-1) \pm \delta x)^2$$

$$= x_i^2(n-1) \pm 2x_i(n-1)\delta x + (\delta x)^2 \qquad (2.17)$$

The mean of this quantity, $<x^2(n)>$, may be expressed as

$$<x^2(n)> = \sum_{i=1}^{N} x_i^2(n)$$

$$= 1/N \sum_{i=1}^{N} \{x_i^2(n-1) \pm 2x_i(n-1)\delta x + (\delta x)^2\}$$

$$= <x^2(n-1)> + (\delta x)^2 \qquad (2.18)$$

If we define $x_i(0)$ as the origin, $<x^2(1)> = (\delta x)^2$, $<x^2(2)> = 2(\delta x)^2$ and $<x^2(n)> = n(\delta x)^2$. We may transform this expression to a function in time by noting that it takes a time t to make n steps whose mean duration is τ. Therefore

$$<x^2(t)> = (t/\tau)(\delta x)^2 = ((\delta x)^2/\tau) t \qquad (2.19)$$

This is more commonly written as

$$<x^2(t)> = 2\Gamma t \qquad (2.20)$$

where Γ is the diffusion constant, defined as $(\delta x)^2/2\tau$. It would appear that we have arbitrarily introduced a factor of 2 as both a numerator and denominator in Eqn 2.20; we shall see below why this complication was introduced. We represent this result in Fig. 2.9 as the change in distribution of particles with

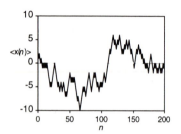

Fig. 2.8 Displacement x for a single one-dimensional random walk with n unit steps.

time, having released them all at $x=0$. The width of the distribution changes with t as $t^{1/2}$. Suppose now that the random motion occurs in a box of particles which has a concentration gradient in the direction x. We show such a box in Fig. 2.9. Intuitively, we would expect more particles to move from regions of higher concentration to those of lower concentration. Let us try to quantify this. The concentration is taken to fall with x from left to right. The number of particles moving from left to right, $N(L-R)$, over a distance δx is given by

$$N(L-R) = -[N(x+\delta x) - N(x)]/2 \qquad (2.21)$$

The flux of particles J_x is the number of particles moving per unit area, per unit time and if we define it as positive if it is from left to right then

$$J_x = N(L-R)/(A\tau)$$

$$= -[N(x+\delta x) - N(x)]/(2A\tau) \qquad (2.22)$$

Recall that $\Gamma = (\delta x)^2/2\tau$, so if we substitute $(\delta x)^2/\Gamma$ for 2τ we find

$$J_x = -\Gamma[N(x+\delta x)/A\delta x - N(x)/A\delta x] \qquad (2.23)$$

Now, $N(x)/A\delta x$ is the number of particles per unit volume and this is equivalent to the concentration, $u(x)$, so

$$J_x = -\Gamma[u(x+\delta x) - u(x)]/\delta x \qquad (2.24)$$

If δx is very small then the above expression may be replaced by a derivative:

$$J_x = -\Gamma \partial u/\partial x \quad \text{as} \quad \delta x \to 0 \qquad (2.25)$$

This is known as Fick's First Law of diffusion. We now see that the factor of 2 introduced in the definition of Γ (eqn 2.20) has the function of making this expression tidier. In three dimensions, eqn 2.25 becomes:

$$\boldsymbol{J} = \Gamma \nabla u \qquad (2.26)$$

where ∇u is called the gradient in u, which for Cartesian coordinates in three dimensions is defined as

$$\nabla = (\hat{\boldsymbol{i}}\, \partial/\partial x + \hat{\boldsymbol{j}}\, \partial/\partial y + \hat{\boldsymbol{k}}\, \partial/\partial z) \qquad (2.27)$$

where $\hat{\boldsymbol{i}}$, $\hat{\boldsymbol{j}}$, and $\hat{\boldsymbol{k}}$ are unit vectors along x, y, and z.

As long as there is a concentration gradient, the concentration at a particular position will change with time. We may quantify this change by returning to one dimension and considering the flux in and out of a section of cross-sectional area A and length δx in the box of Fig. 2.10.

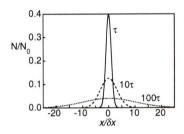

Fig. 2.9 Distribution of particles released at $x=0$ and $t=0$ and allowed to perform a random walk along $\pm x$ for the times marked on the curves.

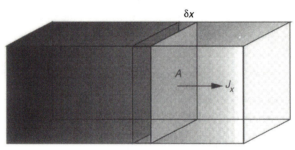

Fig. 2.10 Diffusion in a box of cross-sectional area *A* in which the particle density *N*(*x*)
falls with x from left to right.

In a short time δt the number of particles moving in or out of the box through
the end at position x is $J_x A \delta t$, and the net flux at the other end, at position ($x +$
δx), is $J_x(x+\delta x)A\delta t$. The net change in particles within the box, δN, is then

$$\delta N = -A\delta t(J_x(x+\delta x)-J_x(x)) \tag{2.28}$$

The change in the number of particles per unit volume, which is the change
in concentration, is given by

$$\delta N/(A\delta x)= \delta u = - \delta t(J_x(x+\delta x) - J_x(x))/\delta x \tag{2.29}$$

As $\delta t \rightarrow 0$ and $\delta x \rightarrow 0$ we find the limit

$$\partial u/\partial t = -\partial J_x/\partial x \tag{2.30}$$

After substituting for J_x with Fick's First Law (2.25, 2.26), we obtain Fick's
Second Law of diffusion:

$$\partial u/\partial t = \Gamma \partial^2 u/\partial x^2 \tag{2.31}$$

In three dimensions this is expressed as

$$\partial u/\partial t = \Gamma \nabla^2 u \tag{2.32}$$

where ∇^2 is the three-dimensional Laplacian operator, defined by

$$\nabla^2 = (\partial^2/\partial x^2 + \partial^2/\partial y^2 + \partial^2/\partial z^2) \tag{2.33}$$

The diffusion equation (2.32) may be solved explicitly if we supply one
boundary condition for time, and two for space – that is if we define the state
of the system at one point in time and two points in space. One solution to
which we shall return concerns the one-dimensional case where all N_0
particles start at $x=0$ at $t=0$ and then diffuse from the origin. The number N
at x is then given by:

$$N = \frac{N_0}{\sqrt{4\pi\Gamma t}}\exp(-x^2/4\Gamma t) \tag{2.34}$$

N/N_0 is the probability distribution for a random walk in one-dimension. This was the equation we used to generate the curves in Fig. 2.9.

We shall not dwell on general solutions to the diffusion equation but rather take note of its form when DLA occurs. In the electrochemical growth experiment, as well as the computer simulation of DLA, it may be assumed that the rate of change of the growth front is very slow compared with the rate of particle diffusion. Consequently, we assume that the following conservation law holds

$$\partial u / \partial t = 0 \qquad (2.35)$$

The diffusion equation (2.32) then becomes

$$\nabla^2 u = 0 \qquad (2.36)$$

This is called the Laplace equation. The rate v_n at which the normal to the growth front advances is proportional to the flux of particles to the surface which is given by eqn 2.26:

$$v_n = \hat{\boldsymbol{n}} \, \Gamma . \nabla u \qquad (2.37)$$

where $\boldsymbol{n}$ is a unit vector normal to the growth front.

Equations 2.36 and 2.37 have been derived for a system in which the driving potential arises from a gradient in concentration; analogous expressions exist for systems in which the driving potential is provided by gradients in temperature, hydrostatic pressure, or electrostatic potential and in which a conservation law analogous to eqn 2.35 applies. However, before we look at the solution of the diffusion equations in these and other cases, there is one more ingredient that must be added before we have a general recipe for fractal growth in real materials: we must also consider the change in the energy of the interface between the growing cluster and its environment.

2.6 Interfacial tension

The creation of a small element of surface $\delta\sigma$ between two phases requires a small amount of work δw, this is given by:

$$\delta w = \gamma \, \delta\sigma \qquad (2.38)$$

where γ is the surface tension of the material. Suppose the two phases are two fluids, and one fluid exists as a spherical bubble of radius r and internal pressure p_{in} embedded in the second fluid at a pressure p_{out}. This is the state of affairs when a bubble of vapour forms when a liquid is heated above the boiling point. At equilibrium, the surface tension, which acts to minimise the area of the interface, acts against the force exerted due to the pressure difference. The force exerted by a pressure is given by the pressure multiplied by the area over which it applies, which in this case is $4\pi r^2$. The force that arises through surface tension is the rate at which the energy of the interface increases with the radius of the bubble. The energy of the interface is the product of the area of the interface multiplied by the surface tension, equal to $4\pi r^2 \gamma$. Thus, the surface tension provides a force against the increase of the radius of $8\pi r\gamma$ and the equilibrium condition becomes:

$$4\pi r^2 p_{in} = 4\pi r^2 p_{out} + 8\pi r\gamma \tag{2.39}$$

This gives us

$$p_{in} = p_{out} + 2\gamma/r \tag{2.40}$$

This is also called the Laplace equation. This could cause confusion, so in future we shall always take 'Laplace equation' to be that defined in eqn 2.36.

Surface tension may also influence boiling and melting points. Consider a liquid that is cooled through its melting point T_m. We might envisage the first formation of solid as nucleation of spheres of radius r. This is discouraged by the surface tension $4\pi r^2 \gamma$, and it is possible to supercool the solution by ΔT. This will lead to a change in free energy of fusion ΔG_m of magnitude $\Delta T \Delta S_m$, where the entropy change associated with fusion, ΔS_m, may be expressed in terms of the latent heat of fusion, L, as L/T_m. Therefore, the overall change in free energy of fusion that arises through the surface tension is

$$\Delta G_m = -(4\pi r^3/3)(L\Delta T/T_m) + 4\pi r^2 \gamma \tag{2.41}$$

The critical radius $r*$ below which nucleation cannot occur for a particular value of ΔT is then given by finding the minimum in ΔG_m with respect to r, which differentiation of 2.41 shows to be

$$r* = 2\gamma T_m/L\Delta T \tag{2.42}$$

This is one form of the Gibbs–Thompson equation. One interpretation of this equation is that as the radius of curvature of the interface between two phases gets smaller, so the melting point becomes lower.

Surface tension provides a moderating influence on fractal growth. It discourages the formation of very fine tips in the growth front because high curvature (small values of $1/r$) leads to a high interfacial energy. During the growth of colloidal aggregates such effects may be negligible, but during the growth of solids from supercooled or supersaturated solutions, they may provide the controlling influence. However, the most dramatic manifestation of this effect is found in the flow of a fluid when it is injected under pressure into a second, less viscous fluid.

2.7 Viscous fingering

Take two glass plates measuring approximately 150 mm square and 6 mm thick, and drill a hole of diameter 2–3 mm in the centre of one of them. If this is impractical, you may find it easier to use perspex plates. Glue the end of a syringe or the nozzle from a source of compressed gas into the hole so that it makes a seal that will withstand considerable pressure. The other plate is placed on a tray on a flat surface and smeared with a viscous liquid such as glycerol ($CH_2(OH)CH_2(OH)CH_2(OH)$). The first plate is then placed on top of the second to make a sandwich. The two plates are kept a small distance apart by inserting thin metal sheets such as the feeler gauges used to check the gap on a car's spark plugs. Finally, the two plates are clamped together using the type of clips that artists use to clamp their working surface to a board, and either a gas or a less viscous fluid is injected through the hole in the upper plate. You should see fingers of the less viscous fluid reaching into the more viscous fluid, forcing some of it out of the sandwich, into the tray. The experimental set-up is shown in Fig. 2.11 together with a typical pattern.

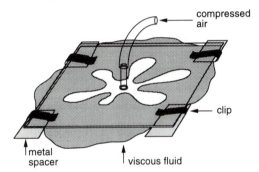

Fig. 2.11 Experimental set-up for the observation of viscous fingering.

Let us first assume that there is negligible interfacial tension. This can be arranged if we choose the fluids carefully. The rate v_n at which the boundary between the fluids moves is proportional to the gradient in pressure P

$$v_n = -M\nabla P \qquad (2.43)$$

where M is the fluid mobility, defined as the ratio of the permeability k to the viscosity μ. If the separation of the plates is b, the permeability is $b^2/12$. If fluid 2 is incompressible, which is a good approximation for most fluids under these conditions, then v_n will not change with position within fluid 2, i.e. $\nabla v_n = 0$. Therefore, using eqn 2.43 we may write

$$\nabla^2 P = 0 \qquad (2.44)$$

which is the Laplace equation again. As with DLA, there is a simple intuitive interpretation of the growth process. Fluctuations in the growth front will produce regions of positive and negative curvature, and eqn 2.43 tells us that whenever fluid 2 protrudes into fluid 1, giving a higher value of ∇P, the value of v_n is higher and the protuberance grows faster than the rest of the front. The growth front is said to be unstable with respect to viscous fingering. This arrangement of plates and fluids described at the start of the section is a form of what is called a radial Hele–Shaw Cell after the Victorian engineer who designed it. With most combinations of fluids the interfacial tension opposes the growth of instabilities because it disfavours regions of high curvature. The width of a finger must then be greater than a critical width λ_c given by

$$\lambda_c = \pi b \sqrt{(\gamma/3 \mu v_n)} \tag{2.45}$$

The fastest-growing fingers are wider than this and dominate the form of the front: their width λ_m is $\sqrt{3}$ greater than λ_c.

Fine fingers are favoured by low interfacial tension, high viscosity, and fast growth. In general, the growth of viscous fingering patterns with fine features requires relatively high pressures and this may be beyond the capabilities of the simple apparatus described above, particularly if the cell uses perspex plates. In the limit of no interfacial tension, eqn 2.45 implies that there is no short-range cut-off to the fineness of the fingers, and any noise in the system may be amplified to produce fractal patterns. This is exemplified by a cell in which the viscous fluid is an aqueous colloidal suspension of a material such as clay, and the less viscous fluid is water, as depicted in Fig. 2.1(b). The degree of noise may be increased by conducting the fluid displacement in a porous medium. The average fluid flow rate will still depend on the pressure gradient in the manner given in eqn 2.43, with k taken to be the mean permeability of the medium. However, there are now random fluctuations in the local pore size, giving random fluctuations in the effective value of b in eqn 2.45, and a wider spread of finger widths and branching points as invasion proceeds. We can modify our growth cell to include this effect by introducing a random collection of fine polystyrene balls between the glass sheets so that flow occurs through the randomly distributed interstices.

The fractal properties of viscous fingering are important in the extraction of oil from porous rocks by the injection of water. It is of great commercial interest to be able to predict how efficient this is as a recovery method. The tendency to form fractal viscous fingering patterns suggests that such extraction methods are very inefficient, with only a small proportion of oil being displaced before the water breaks through to the point from which the oil is pumped. The efficiency of the extraction may be improved by adding substances to the water which increase the tension of the oil-water interface or increase the viscosity of the water.

The viscous fingering problem may be modified in a significant manner by allowing reaction to occur between the porous medium and the invading fluid. Such a process controls the development of interlinked cave systems in limestone, and the modification of porous, oil-bearing rock when it is injected with an acid to facilitate the extraction of the oil. If we assume that reaction between the acid and the rock is limited by the rate at which the dissolution products can be carried away, then the rate of dissolution is proportional to the local velocity of the fluid. We shall see other examples of diffusion limited reaction in chapter 4. Faster dissolution leads to greater local permeability which in turn leads to greater local flow. This form of feed-back provides the necessary conditions for Laplacian growth. A simple experiment which demonstrates such a process takes a layer of plaster of

Paris (CaSO$_4$.0.5H$_2$O) held between two glass plates as the reactive porous medium, and water as the corrosive fluid. The reaction is initiated by injecting the water through the centre of one of the glass plates, stopped after several hours and the eroded plaster then used as a mould to cast a permanent record of the dissolution pattern in a low-melting metal alloy such as Woods metal. The remaining plaster is dissolved away, and the metal found to have a fractal dimension of 1.6±0.1 over about 3 decades of length-scale.

2.8 Dendritic growth

When we described the electrodeposition of polypyrrole in Section 2.2 we commented on the fact that as the potential of the anode was increased, the polymer changed from a compact shape through a branched form with some degree of regularity in the branching angles, to growth characteristic of DLA. This intermediate type of structure is called 'dendritic' (*dendron* means 'tree' in ancient Greek). It is a common term in mineralogy and metallurgy, and refers to fern-like growths of rocks and elemental metals, and the form of many intermetallic alloys.

(a)

(b)

(c)

(d)

Fig. 2.12 Natural dendrites: (a) pyrolusite in limestone; (b) Jack Frost at work in a Canadian winter; (c) moss agate; (d) Al$_6$Fe grown by quenching a molten alloy of 10 per cent Fe in Al.

In Fig. 2.12(a) we show a very common type of dendrite, found in fissures in many sedimentary rocks such as sandstone and limestone. These materials present a layered, porous medium to an invading solution rich in an impurity ion. Common examples are deposits of a type of MnO$_2$ called pyrolusite, which is formed when a solution rich in manganese salts infiltrates sediments under oxidizing conditions. As the sediment dries out, the fractal invasion front is highlighted as the black, fern-like deposits shown the figure. A similar process might be responsible for the beautiful patterns found in moss agate (Fig. 2.12(b)). In that case the host material is a microcrystalline form of silica called chalcedony and is formed from a colloidal aqueous solution trapped in a non-porous shell. Solutions rich in manganese or iron salts invade this medium, staining it green or black in characteristic patterns.

These dendrites are random fractals and may be described well using DLA models. For the remainder of this book we shall reserve the term 'dendritic growth' for a process which reveals some underlying symmetry in the binding process. Snowflakes provide a good natural example of this phenomenon (Fig. 2.12(b)). Although their growth from a vapour involves aggregation by random walks, there is an anisotropy in the binding process at a molecular level which leads to a macroscopic symmetry. While no two snow-flakes may be identical, their general form is very characteristic. Intermetallic alloys and salts grown from melts and supersaturated solutions may also display highly branched structures whose mass scales with length as a fractal would, but which clearly have some overall symmetry (Fig. 2.12(d)). We shall consider first how growth of such solids may become unstable in a manner analogous to the instability of viscous fingering. We shall then consider how anisotropy in diffusion processes leads to dendritic growth in a wider variety of processes.

Stable and unstable growth from melts and supersaturated solutions

Consider what happens when we cool the walls of a box full of a pure liquid by an amount Δ below its melting point T_m. The nucleation commonly occurs at the walls of the vessel, and the front will move smoothly into the liquid as depicted in Fig. 2.13(a). This is called stable growth. The rate v_n at which the normal to the front moves into the melt is determined by the rate at which the latent heat of fusion, L, may be conducted away. The rate of diffusion of heat into the solid, J_S, is given by

$$J_S = C_p^S \, \Gamma^S \nabla T^S \tag{2.46}$$

where C_p^S and Γ^S are the heat capacity per unit volume and the thermal diffusion coefficient of the solid phase respectively. A similar expression governs the rate of diffusion into the liquid, J^L. The rate of generation of heat per unit volume is Lv_n, so for steady-state growth v_n is given by

$$Lv_n = [\, C_p^S \, \Gamma^S \nabla T^S - C_p^L \, \Gamma^L \nabla T^L \,]. \, \hat{\boldsymbol{n}} \tag{2.47}$$

where $\hat{\boldsymbol{n}}$ is a unit vector normal to the growth front. Suppose that nucleation had occurred in the centre of the box. The growth front would then move into supercooled liquid. Any protuberance in the front reaches into the melt at a lower temperature and therefore grows faster.

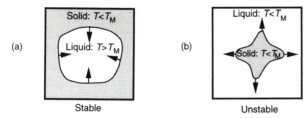

(a) Stable

(b) Unstable

Fig. 2.13 (a) Stable and (b) unstable growth of solid (the shaded region) from a liquid which is cooled below its melting point by lowering the temperature of the walls of the container.

This form of unstable growth is known as a Mullins–Sekerka instability after the first people to provide an analytic mathematical description of the phenomenon, and is depicted in Fig. 2.13(b). Let us consider this in more detail, using growth in one direction, x, for simplicity.

The driving force for growth is provided by the supercooling $\Delta \equiv T - T_m$. It is common to define a dimensionless diffusion field u', called the undercooling, as follows:

$$u' = C_p \Delta / L \tag{2.48}$$

Note the similarity between eqn 2.47 and eqn 2.37 when we make the substitution for u'. The flow of heat in the system obeys a diffusion equation of the form 2.31. Under steady–state conditions, v_n is constant with time and it is convenient to transform our coordinates so that x is expressed relative to the growth front as x' where $x' = x - v_n t$. The transformed equation is then

$$\Gamma \partial^2 u / \partial x'^2 = -v_n \partial u / \partial x' \tag{2.49}$$

A solution of this equation for the liquid phase is

$$u = \exp(-x' v_n / \Gamma) - 1 \tag{2.50}$$

The temperature of the interface, T_i, is the melting point corrected for the effects of interfacial tension according to the Gibbs–Thompson relation (Eqn 2.42)

$$T_i = T_m(1 - \gamma \kappa / L) \tag{2.51}$$

where κ is the mean inverse radius of the growing front, equal to $(1/r_1 + 1/r_2)$, where r_1 and r_2 are the principal radii of the front. This is sometimes written in terms of a quantity d_0 called the diffusion length:

$$u = -d_0 \kappa \tag{2.52}$$

where $d_0 = \gamma T_m C_p / L^2$. We plot the solution of the heat flow equation (2.50) in Fig. 2.14, illustrating the fall in temperature away from the growth front. This treatment assumes steady–state growth conditions, which appears to conflict with the observation that the growth front may be unstable with respect to the development of bumps. So long as the rate at which a portion of the front moves is small compared with Γ, then the growth conditions may relax rapidly compared to the development of the front, and eqn 2.50 continues to apply. This occurs when the diffusion field is small: $|u|\ll1$. We note in passing that eqn 2.49 then approximates to the Laplace equation (2.36). Any protuberance in the growth front now causes the thermal gradient at that point to increase, as depicted schematically in Fig. 2.14, and this in turn produces faster growth locally.

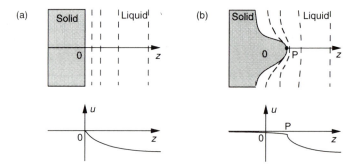

Fig. 2.14 Isotherms for (a) a flat and (b) a distorted front for a solid growing into a cooled liquid: temperature falls from left to right. The figure could also represent the growth of a solid from a supersaturated solution with the contours indicating the increase in concentration from left to right.

As we observed with the viscous fingering, the development of such tips is impeded by an increase in surface tension. A balance is struck between the two competing influences such that the interface develops a periodic deformation whose wavelength λ is constrained by surface tension to the minimum value

$$\lambda = 2\pi\alpha\sqrt{2\Gamma d_0/v_n} \qquad (2.53)$$

where α is a dimensionless constant defined as

$$[(1+\beta)/2]^{1/2} \qquad (2.54)$$

and β is $C_p^S\,\Gamma^S/C_p^L\,\Gamma^L$.

A similar instability may develop when a solute solidifies from a supersaturated solution. This could be an aqueous solution of an ionic salt or a binary metallic alloy. In either case we may recast the thermal diffusion problem above in terms of diffusion in a concentration field. Suppose the concentration of solute in the solid phase is C_s (which is equal to the density for the formation of a pure salt), and the equilibrium concentration of solute in solution at that temperature is C_{eq}, then the solidification is driven by holding the concentration of solute in solution at a much higher concentration. Clearly, the concentration C of solute at an arbitrary position, x will rise from the solidification front to some limiting value C_∞ a great distance away. The driving force for diffusion may be expressed as the dimensionless quantity u'

$$u' = (C - C_{eq})/(C_s - C_{eq}) \qquad (2.55)$$

which is similar to eqn 2.48. The flow of solute and the motion of the front are governed by a diffusion equation similar to eqn 2.49 and if the front moves slowly relative to Γ, which is the case when the driving force for diffusion, $(C_\infty - C_{eq})/(C_s - C_{eq})$, is small, then u' depends on x in a manner analogous to 2.50. Any protuberance in the growth front takes the solid into a region of higher supersaturation and hence faster growth. As with the thermal solidification problem, the interfacial tension sets a lower-limit to the length–scale of the instabilities so that the minimum wavelength for an unstable distortion is given by eqn 2.53, in which the chemical diffusion length d_0 may be expressed as

$$d_0 = \gamma/(C_s - C_{eq})^2 (\delta\mu/\delta c) \qquad (2.56)$$

Growth in either a thermal or a concentration diffusion field is exemplified by the solidification of succinonitrile ($NCCH_2CH_2CN$) from a supercooled melt or NH_4Br from a supersaturated aqueous solution. NH_4Br may be crystallized to form dense, cubic crystals under conditions of thermodynamic control. When grown from supersaturated solution four side-branches grow in the <100> directions. It is clear that some form of four-fold symmetry in the binding probability for particles attaching themselves to the growth front leads to an overall symmetry in the solid. These principal branches are themselves unstable with respect to further branching. The growing tips of radius ρ develop an instability in the form of secondary tips that start to form several multiples of ρ from the tip. As the principal branches continue to develop, the position of the first side-branch remains fixed relative to the tip, but its amplitude increases and it is joined by further side–branches further from the tip. This is illustrated in Fig. 2.15. It is clear that although the size and spacing of the side-branches have some regularity, as embodied in eqn 2.53, as well as a regular direction, consistently perpendicular to the principal axis of the main branch, the growth also has a noisy component.

Some success has been found in the computer simulation of dendritic growth patterns with a particular symmetry. The lattice on which the random walk occurs may impose such a symmetry. This may not be apparent for relatively small simulations where the noise provided by the randomness of the walk dominates the growth and masks the underlying symmetry; larger simulations may appear to adopt of the symmetry of the host lattice as is apparent when Fig. 2.16(a), which has $3.82 . 10^6$ sites, is compared with Fig. 2.5(b), with $5 . 10^4$ sites. The noise in the growth process may also be reduced by modifying the DLA simulation such that a particle only attaches itself to the growing cluster after it has touched it m times. As m is increased so the noise is suppressed. Figure 2.16(b) shows a simulation for a square lattice which is much smaller ($5 . 10^5$ sites) than that of Fig. 2.16(a), yet appears more symmetric. The effect of making m very large is demonstrated

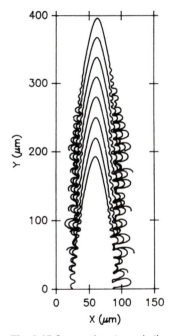

Fig. 2.15 Successive stages in the growth of a branch of a NH_4Br dendrite from supersaturated aqueous solution. The contours represent successive positions of the growth front taken at 10 s intervals from botom to top using an optical microscope.

in Fig. 2.16(c) in which a simulation is performed on a hexagonal lattice for 10^4 sites with $m = 500$.

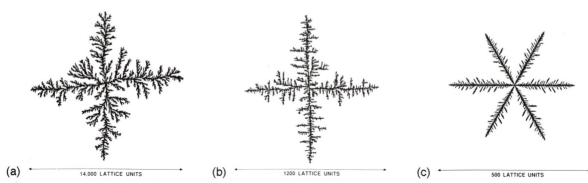

(a) 14,000 LATTICE UNITS (b) 1200 LATTICE UNITS (c) 500 LATTICE UNITS

Fig. 2.16 (a) Dendritic growth produced by computer simulations in which an overall symmetry is imposed by the square symmetry of the lattice for long simulations ($3.82 . 10^6$ sites). Simulations on (b) a square lattice with $5 . 10^4$ sites and (c) a hexagonal lattice with 10^4 sites with noise reduction factors $m = 3$ and 500 respectively.

A radial Hele–Shaw cell may be modified to produce dendritic growth. If one of the glass plates is etched with a fine triangular lattice, the anisotropy in the pressure field leads to viscous fingering that resembles a snow–flake. However, it is still not clear what is the connection between the symmetry of the bonding at an atomic or a molecular level, and the symmetry of the overall structure. We are a long way from being able to predict how a snow–flake will grow with a knowledge of the appropriate intermolecular potentials.

Solidification processes may produce random fractals if the anisotropy varies randomly with position, just as a random distribution of pressure fields produced fractal invasion fronts when viscous fingering was performed in a random, porous medium. NH_4Cl may be grown dendritically in the manner we described above for NH_4Br. If the growth is induced in a thin glass cell in which one of the glass plates is abraded randomly, the random branching structure of Fig. 2.17 is observed.

A random environment may also be produced in the crystallization of solids from amorphous forms. $GeSe_2$ may be deposited as an amorphous thin film from a vapour and then annealed below its glass transition temperature to induce crystallization in the form of fractal clusters. Digitization and image analysis of transmission electron micrographs of the clusters yielded $D \approx 1.69$, close to the value for DLA.

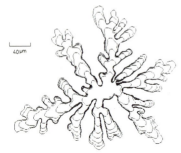

40 μm

Fig. 2.17 Random solidification front produced in NH_4Cl grown from supersaturated solution in a flat cell with a randomly abraded surface.

2.9 Summary

There is a wide variety of objects whose surface area or distribution of mass or voids shows fractal scaling. These include the following.

(a) Diffusion-limited aggregates grown electrochemically or from a sol or aerosol or on a computer.

(b)   Viscous fingers when interfacial tension is negligible and growth occurs in a porous medium.

(c)   Solids grown from supercooled or supersaturated solutions by a diffusion-limited process in a random environment.

The growth of all of these objects depends on the diffusion of particles or fluid or heat according to the Laplace equation ($\nabla^2 u = 0$), which is the form adopted by the general diffusion equation when $\partial u/\partial t = 0$. u is proportional to concentration, pressure, or heat and diffusion is driven by a gradient in these properties. The velocity of the growth front is proportional to ∇u.

Instability of the growth front with respect to the development of protuberances and the neglect of depressions may be limited by surface tension for viscous fingering and solidification. This sets a lower limit to the curvature of a growth front. When there is also an anisotropy present which favours growth in certain directions, dendritic growth may occur: the growth then shows some regularity in the directions taken by the branching pattern. Anisotropy may be provided either by the growth environment, or the binding probability at the growth front in a solidification process. The balance between random and dendritic growth is summarized in Fig. 2.18.

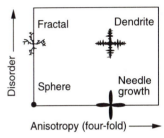

Fig. 2.18 Schematic illustration of the development of random fractal or dendritic forms as either the degree of randomness or the anisotropy in the growth process is increased, starting from a spherical form.

Further reading

In addition to the relevant chapters in the book edited by Avnir, and the book by J. Feder, you may find the following books and articles useful.

Sander, L.M. (1986). Fractal growth processes. *Nature*, **322** 789.
J. Langer, J. (1980). Instabilities and pattern formation in crystal growth. *Review of Modern Physics*, **52** 1.
Van Straaten, L.M.J.U. 1978). Dendrites. *Journal of the Geological Society of London*, **135** 137.
Stanley, H.E. and Ostrowsky, N. (eds). (1986). *On growth and form*. NATO ASI Series 100E, Kluwer, Dordrecht.
Stanley, H.E. and Ostrowsky, N. (eds). (1988). *Random fluctuations and pattern growth*. NATO ASI Series 157E, Kluwer, Dordrecht.
Vicsek, T. (1989). *Fractal growth phenomena*. World Scientific, London.
Kessler, D.A., Koplik, J., and Levine, H. (1988). Pattern selection in fingered growth phenomena. *Advances in Physics*, **37** 255.

Problems

2.1. In the Witten and Saunder model for particle aggregation, the probability $u(r, n+1)$ that a site at a position r on the grid gains a particle after $n+1$ steps of a random walk is given by the mean of the probabilities that the c neighbouring sites are occupied in the previous step i.e.

$$u(r, n+1) = \frac{1}{c} \sum_{\alpha=1}^{c} u(r+\alpha, n) \qquad (2.57)$$

where α is a vector to the c nearest neighbours. Show that the diffusion of particles on the grid is controlled by a discrete version of the diffusion equation (2.32) and that the growth of the cluster is controlled by a discrete version of eqn (2.37), given that $u(r,n)=0$ when r corresponds to the position of the growth front.

2.2. The mean-square displacement from the starting point of a particle that executes a random walk for a time t in just one dimension is given by eqn 2.20. Show how this expression may be generalised to E dimensions. Calculate the number of sites that are enclosed in the line, or area or volume swept out by the walk, and the average time spent at each site if they are separated by a distance α along any orthogonal axis. What is the likelihood of the particle returning to the origin when $E = 1$ and when $E = 3$?

2.3. The branching pattern formed in air when lightening strikes looks as though it might have a fractal structure. This phenomenon is an example of dielectric breakdown, and occurs when gas particles in the atmosphere are ionised in a catastrophic fashion in a high electrical potential, ϕ. A similar type of pattern may be reproduced in the laboratory in a more manageable fashion by applying a high electrical potential to a compressed gas such as SF_6, which has a high ionisation threshold. When the electrodes take the form of a ring and a central point, the discharge pattern resembles the DLA patterns of Fig. 2.1 with the branches representing the conducting ionised region. The development of the pattern is controlled by Poisson's equation:

$$\nabla^2 \phi = Q \tag{2.58}$$

where Q is the charge density, which is zero in the insulating region. In the conducting region ϕ is zero. Point out the analogies between this type of growth and both viscous fingering and the dendritic growth processes.

The branches of ionised gas have a constant width, so the increase in the mass of the object with radius from the centre reflects the number $n(r)$ of branches as a function of r. If we assume that the growth has the same fractal dimension as DLA for $E = 2$, state how $n(r)$ will scale with r.

3 Characterising fractals

3.1 Introduction

In this chapter we describe three methods for the determination of the spatial distribution of mass in fractal and non-fractal objects. First, we return briefly to the analysis of photographs of fractal objects using techniques such as the box-counting method which we introduced in Chapter 1. We then consider small-angle scattering measurements in some detail because of their widespread use as a means of measuring the distribution of mass in colloids and polymers. Finally we describe experiments in which the rate at which energy is transferred between molecules adsorbed on a fractal surface provides an estimate of the surface or pore fractal dimension.

There are other properties of a material that may be used to provide an estimate for D, notably the measurement of the coverage of a rough surface with molecules of a known area and the rate at which a particle or energy diffuses through a fractal medium. We prefer to leave these subjects to Chapter 4 which is concerned with the influence of D on some of the physical and chemical properties of matter.

3.2 Image analysis

In Chapter 1 we described how the fractal dimension of an object could be estimated from a photograph by measuring the length of the boundary, or the area of the object with a range of resolutions and then applying the suitable fractal scaling law. The walking technique provides an estimate for D of a boundary with $D_T = 1$, and the box-counting technique for boundaries or surfaces with D_T equal to 1 or 2. Manual execution of these procedures may be laborious, so various algorithms have been developed for the manipulation of images stored in the memory of a computer.

First we need to obtain a photograph. Many of the objects we have described in the preceding chapters are microscopic in scale, and we need to take pictures with a conventional optical microscope whose resolution is of the order of $10\,\mu m$, or with various forms of electron microscope for objects down to atomic length-scales. The image may then be transferred to the computer using an optical scanner or a video camera connected to a port. Both methods convert the image to an array of squares called pixels, and the computer memory holds these as a matrix whose elements take values that reflect the colour or darkness of the corresponding part of the original image. Usually, the pixel is merely recorded as being occupied or unoccupied, depending on the original image. The edge of a pixel typically corresponds to about $100\,\mu m$ in the image that has been scanned or video-recorded.

conventionally requires a nuclear reactor source, monochromation through Bragg reflection, and detection with a form of Geiger counter. Small-angle light scattering (SALS) employs a laser as a light source, slits, filters, and lenses to direct the required radiation onto the sample, and a photomultiplier as a detector.

The different techniques have different working ranges of wavelength, and hence different accessible ranges of d; these are summarized in Table 3.1.

A second important difference between the techniques stems from the

Table 3.1 Different characteristics of different scattering techniques. The table gives the range of wavelength and scattering vector Q provided by small-angle scattering of X-rays (SAXS), neutrons (SANS), or light (SALS). It also indicates the microscopic origin of the scattering

Radiation	Source of scattering potential	Wavelength range (Å)	Q range (Å^{-1})	Resolution (Å)
X-rays	Electrons	1 – 4	10^{-2} – 15	0.5 – 500
Neutrons	Nuclei and magnetic moments	1 – 30	10^{-3}–15	0.5 – 5000
Light	Refractive index/ dielectric constant	4000 – 6000	5.10^{-5} – 3.10^{-3}	2000 – 100,000

microscopic nature of the scattering. If a parallel, monochromatic beam of visible light enters the medium from, say, a vacuum, its speed may change and it may be refracted, but the rays comprising the beam will remain parallel. It will be scattered if it experiences inhomogeneities in the refractive index whose size is comparable to or larger than the wavelength. We say that there is a *contrast* between the scatterers and the medium. In the case of X-rays, scattering is from inhomogeneities in electron density and the contrast scales with atomic weight Z as Z^2. Neutrons are scattered by the nucleus and the magnetic moment of an atom and the scattering strength varies with Z in a more complex manner: large fluctuations are superposed on a shallow increase of scattering ability with Z. In particular, the scattering ability of hydrogen and deuterium are quite different and the mean scattering ability of an aqueous or organic solvent containing H or D may be controlled by altering the isotopic ratio so that it is equal to that of the particles in solution. The medium would then appear homogeneous and scattering will not occur. This is very useful when there is a variety of particles in solution, or when they are particularly complex. By matching the contrast of the solvent with certain types or parts of the particles and causing them to disappear, the measurements provide information about a subset of the scatterers. An optical analogy for this process is provided by considering a set of coloured glass marbles of different colours which may be embedded in a glass of a particular colour. Only those parts of the marbles which are of a different colour to the host material would then be visible.

In the next two sections we develop scattering theory in terms of the interference of waves, and show how such measurements provide a probe of the distribution of fluctuations in scattering strength. If you are already familiar with this you may wish to skip to eqn 3.22 on page 54.

Interference of waves

We may represent the wave $\psi(x)$ travelling in the direction x by the function

$$\psi(x) = f \cos(2\pi x/\lambda) \qquad (3.2)$$

where the maximum displacement f occurs at $x = 0$ and integral multiples of the wavelength λ. We may displace the positions of the maxima by the angle ϕ, called the phase, which gives us a more general expression for $\psi(x)$:

$$\psi(x) = f \cos(2\pi x/\lambda - \phi)$$

$$= f \cos(kx - \phi) \qquad (3.3)$$

where $k = 2\pi/\lambda$ and is called the wavevector. Suppose now we have two waves, $\psi_i(x)$ and $\psi_j(x)$, with the same wavevector k but different phase and amplitude given by

$$\psi_i(x) = f_i\cos(kx - \phi_i) = f_i(\cos(kx)\cos \phi_i + \sin(kx)\sin \phi_i) \qquad (3.4)$$

and

$$\psi_j(x) = f_j\cos(kx - \phi_j) = f_j(\cos(kx)\cos \phi_j + \sin(kx)\sin \phi_j) \qquad (3.5)$$

The resultant wave, $\psi(x) = \psi_i(x) + \psi_j(x)$, may also be expressed as

$$\psi(x) = f(\cos(kx)\cos \phi + \sin(kx)\sin \phi)$$

$$= f_i(\cos(kx)\cos \phi_i + \sin(kx)\sin \phi_i) +$$

$$f_j(\cos(kx)\cos \phi_j + \sin(kx)\sin\phi_j) \qquad (3.6)$$

If we compare like terms we find

$$f\cos\phi = f_i\cos\phi_i + f_j\cos\phi_j \qquad (3.7)$$

and

$$f\sin\phi = f_i\sin\phi_i + f_j\sin\phi_j \qquad (3.8)$$

Now, $\cos^2\phi + \sin^2\phi = 1$ so

$$f = [(f_i\cos\phi_i + f_j\cos\phi_j)^2 + (f_i\sin\phi_i + f_j\sin\phi_j)^2] \qquad (3.9)$$

and the phase is given by

$$\tan \phi = (f_i\sin\phi_i + f_j\sin\phi_j)/(f_i\cos\phi_i + f_j\cos\phi_j) \qquad (3.10)$$

The individual waves and their sum are depicted pictorially in Fig. 3.4.

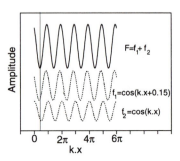

Fig. 3.4 Representation of the superposition of two waves travelling in the direction x with the same wavevector k but with different phase and amplitude. The individual waves are at the bottom, and their sum, also of wavevector k, is at the top.

The addition of waves resembles the addition of vectors. We shall find it more convenient to express the amplitude of the ith wave as a vector f_i

$$f_i = f_i\cos\phi_i + if_i\sin\phi_i \qquad (3.11)$$

where i $= \sqrt{-1}$. This may also be written as

$$f_i = f_i\exp(i\phi_i) \qquad (3.12)$$

The result of adding N such waves of wavevector k is another wave of the same wavevector represented by F where

$$F = \sum_{i=1}^{N} f_i \exp(i\phi_i) \qquad (3.13)$$

The amplitude $F = |F|$ is given by

$$F = \sqrt{A'^2 + B'^2} \qquad (3.14)$$

where

$$A' = \sum_{i=1}^{N} f_i \cos\phi_i \qquad (3.15)$$

$$B' = \sum_{i=1}^{N} f_i \sin\phi_i \qquad (3.16)$$

The phase ϕ is given by

$$\tan\phi = B'/A' \qquad (3.17)$$

If we extend this analysis to waves in three dimensions, k is a vector $\boldsymbol{k}$ whose modulus $|\boldsymbol{k}| = 2\pi/\lambda$.

Interference of scattered waves

In a diffraction experiment, a wave of wavevector $\boldsymbol{k}$ is incident on an object or a collection of objects at positions $\boldsymbol{r}_i$ which scatter it to produce waves of wavevector $\boldsymbol{k'}$ centred at these positions. If the scattering is elastic – that is, if the wavelength does not change on scattering – we may write the spherical scattered waves as

$$\psi_i(r) = (f_i f_0/D)\ \exp(\boldsymbol{k'.r} + \phi_i) \qquad (3.18)$$

where f_0 is the amplitude of the incident wave and f_i is a measure of the efficiency of the ith scatterer. We recall that the efficiency of scattering also depends on the contrast between the particle and the medium in which it is held. At present we assume that we are dealing with a small number of scatterers suspended in a vacuum. D is the distance between $\boldsymbol{r}$ and the centre of the scatterer, and eqn 3.18 reflects the fact that the amplitude of the individual scattered waves decreases as $1/D$. The resultant wave is the superposition of the set of waves $\psi_i(r)$, and may be calculated in the manner described above. First, we need to be able to specify ϕ_i. We assume that the phase change experienced during the scattering event is independent of i. The relative values of ϕ_i now depend on the path differences between the incident and scattered waves as indicated for just two scatterers, 1 and 2, in Fig. 3.5.

We assume that the positions of 1 and 2 are 0 and $\boldsymbol{R}$ respectively and that the scattering process produces the same phase shift at 1 and 2. The relative phase of the two waves travelling in direction $\hat{\boldsymbol{k}}'$ then depends on the additional pathlength $(a + b)$ for waves scattered from 2 relative to those from 1. a is given by the projection of $\boldsymbol{R}$ on the unit incident wavevector $\hat{\boldsymbol{k}}$, which is $-\boldsymbol{R}.\hat{\boldsymbol{k}}$ and b is $+ \boldsymbol{R}.\hat{\boldsymbol{k}}'$. The difference in path length is then $\boldsymbol{R}.(\hat{\boldsymbol{k}}' - \hat{\boldsymbol{k}})$. The phase angle associated with this difference is 2π multiplied by the additional number of cycles the wave goes through, which is $\boldsymbol{R}.(\hat{\boldsymbol{k}}' - \hat{\boldsymbol{k}})/\lambda$. Therefore, the phase of scattering from 2 relative to 1, ϕ_{2-1}, is

$$\phi_{2-1} = 2\pi\boldsymbol{R}.(\hat{\boldsymbol{k}}' - \hat{\boldsymbol{k}})/\lambda = \boldsymbol{R}.(\boldsymbol{k'-k}) \qquad (3.19)$$

The term $(\boldsymbol{k'-k})$ is usually written as $\boldsymbol{Q}$ and is called the scattering vector. It is proportional to the change of momentum of the wave upon scattering and its modulus $Q = |\boldsymbol{Q}|$ is equal to $4\pi\sin\theta/\lambda$. The resultant wave $\psi(\boldsymbol{r})$ is

$$\psi(\boldsymbol{r}) = \psi_1(\boldsymbol{r}) + \psi_2(\boldsymbol{r})$$

$$= (f_0/D)\ \exp(i\boldsymbol{k.r})\{f_1 + f_2\exp(i\boldsymbol{Q.R})\} \qquad (3.20)$$

In general, for N scatterers, the resultant wave $\psi_s(\boldsymbol{r})$ is

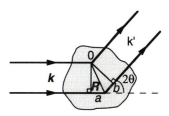

Fig. 3.5 The geometry of scattering from two centres in a particle whose positions are defined as the origin and $\boldsymbol{R}$. The incident and scattered waves have wavevectors $\boldsymbol{k}$ and $\boldsymbol{k'}$ respectively, 2θ is the scattering angle and $(a + b)$ is the extra distance travelled by the wave scattered from the centre at $\boldsymbol{R}$ compared with that at the origin.

$$\psi_s(r) = (f_0 / D) \exp(i\mathbf{k}.\mathbf{r}) \sum_{i=1}^{N} f_i \exp(i\mathbf{Q}.\mathbf{r}_i) \qquad (3.21)$$

The intensity of the scattering into the solid angle $d\Omega$ is proportional to the square of the amplitude of the scattered wave multiplied by D^2, the square of the distance between sample and detector, where D may be taken to be the same as D in eqn 3.20. We usually express this as $I(Q)$, which is proportional to the square of the amplitude of the scattered wave per unit volume of sample, relative to the intensity of the beam that passes straight through the sample. Thus

$$I(\mathbf{Q}) = \frac{1}{V}\left\langle \left| F^*(\mathbf{Q}) \, F(\mathbf{Q}) \right|^2 \right\rangle \qquad (3.22)$$

where

$$F(\mathbf{Q}) = \sum_{i=1}^{N} f_i \exp(i\mathbf{Q}.\mathbf{r}_i) \qquad (3.23)$$

In many cases it is more appropriate to consider the distribution of scattering power to be continuous in space, rather than discrete, and replace the scattering strength of an individual scatterer, f_i, by a volume element dV multiplied by the scattering density $\rho(r)$: $f_i = \rho(r)dV$. The sum in eqn 3.23 then becomes an integral over volume

$$F(\mathbf{Q}) = \int_V \rho(\mathbf{r}) \exp(i\mathbf{Q}.\mathbf{r}) \, dV \qquad (3.24)$$

$F(Q)$ is the spatial Fourier transform of the scattering density. This general principle is the foundation of all diffraction theory. To progress, we need to find a suitable expression for $\rho(r)$. Let us begin by considering the simplest case – that of a spherical particle of uniform scattering density. There we may write $\exp(i\mathbf{r}_i.\mathbf{Q}) = \cos(\mathbf{r}_i.\mathbf{Q})$ since sine is an odd function and terms in $\mathbf{r}_i$ are cancelled out by the equal number of terms in $-\mathbf{r}_i$. The spherical symmetry allows us to replace $\mathbf{Q}$ and $\mathbf{r}$ by their modulus, and if we transform to polar coordinates, such that $dV = r^2 \sin\varphi \, d\phi \, d\varphi \, dr$, we find

$$F(Q) = \int_{r=0}^{\infty} \int_{\varphi=0}^{\pi} \int_{\phi=0}^{2\pi} \rho(r)r^2 \cos(Qr\cos\varphi)\sin\varphi \, d\phi \, d\varphi \, dr \qquad (3.25)$$

or

$$F(Q) = 4\pi \int_0^{\infty} \rho(r) \frac{\sin(Qr)}{Qr} r^2 dr \qquad (3.26)$$

Scattering from aggregates

In many of the applications to aggregate materials which we consider below, we should split the sum of eqn 3.23 or the integral of eqn 3.24 into two portions – one for the distribution of scattering density over the n individual particles, and one for the distribution of the n_p particles. Equation 3.22 then becomes

$$I(Q) = \frac{1}{V} \left\langle \left| \sum_{j=1}^{n} f_j \exp(\mathrm{i}Q.\,r_j) \sum_{i=1}^{n_p} \exp(\mathrm{i}Q.\,R_i) \right|^2 \right\rangle \qquad (3.27)$$

where R_i is the mean position of the *i*th particle, and r_j the position within any particle. If we convert this to an integral representation we find

$$I(Q) = (n/V)\, P(Q)\, S(Q) \qquad (3.28)$$

where $P(Q)$ corresponds to the first sum in eqn 3.27, and $S(Q)$ to the second. For centrosymmetric particles we find

$$P(\boldsymbol{Q}) = P(Q) = \left\langle |F(\boldsymbol{Q})|^2 \right\rangle \qquad (3.29)$$

where $F(\boldsymbol{Q})$ is given by eqn 3.26 with the integration performed over the extent of the individual particles. $S(\boldsymbol{Q})$ is equal to $S(Q)$ and is a similar integral over all space with $\rho(r)$ describing the probability of finding another particle at a distance r. $S(Q)$ is usually expressed in terms of the correlation function $G(r)$, defined as

$$G(r) = (1/n) \int_{V} \rho(r')\rho(r + r')\mathrm{d}\, r'$$

$$= \delta(r) + g(r) \qquad (3.30)$$

The delta function takes the value 1 when r is at the origin and 0 otherwise; it expresses the fact that there is a particle at the origin. $g(r)$ is the static pair correlation function and describes the probability that some other particle is a distance r from the origin, with the position of the origin averaged over the sample volume.

$$S(Q) = 1 + \frac{4\pi n}{V} \int_{0}^{\infty} g(r)\, r^2 \, \frac{\sin(Qr)}{Qr}\, \mathrm{d}r$$

$$(3.31)$$

Before we substitute expressions for $g(r)$ appropriate for fractal objects in eqn 3.31 to obtain $I(Q)$, let us look at the general properties of $I(Q)$ as a function of Q with the aid of Fig. 3.6 which shows a solution of fractal aggregates observed with increasing resolution. This could be thought of as the effect of scattering from a colloidal silicate solution with a decreasing wavelength λ. The colloidal particles have a mean radius R and are constructed from elementary particles of radius a.

   At very long wavelengths ($Q \approx 0$), the waves regard the medium they pass through as containing a distribution of point scatterers whose internal structure does not influence their passage. As λ is reduced so that it is smaller than R, the scattering becomes sensitive to the distribution of elementary particles within the cluster: when $Qa \ll 1$ and $QR \gg 1$, small-angle scattering probes the dimension of *mass* fractals. Further reduction of λ such that $Qa \gg 1$ leads to the situation where the radiation regards either the solvent or the silicate particles as homogeneous, and scattering only occurs as the radiation passes between the two. In this regime, it is the *surface* of the particles that is being studied, so $I(Q)$ provides information about the dimension of surface fractals. Finally, as the wavelength approaches the separation of atoms in the particles, diffraction characteristic of the atomic or the molecular structure of the silicate may be observed, but the angles at which diffraction occurs may no longer be called small.

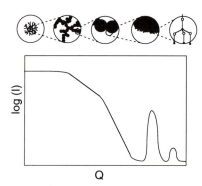

Fig. 3.6 Schematic illustration of the scattering from a colloidal aggregate, showing the aspect of the structure that is probed in the different regions of Q.

Scattering from mass fractals

When Q conforms to the inequalities $QR \gg 1$ and $Qa \ll 1$, $P(Q) \approx 1$ so $I(Q) \approx (n/V) \, S(Q)$. In order to produce an expression for $I(Q)$ we need an expression for $g(r)$. The number of particles $n(r)$ in a sphere of radius r is given by

$$n(r) = (n / V) \int_0^r 4\pi r^2 g(r) \mathrm{d}r \tag{3.32}$$

therefore

$$\mathrm{d}n(r) = (n/V) g(r) \, 4\pi r^2 \mathrm{d}r \tag{3.33}$$

However, if the volume is filled with an object of fractal dimension D made of spheres of radius a, eqn 1.21 tells us that

$$n(r) = p \, (r/a)^D \tag{3.34}$$

Therefore

$$\mathrm{d}n(r) = p \, (D/a)(r/a)^{D-1} \mathrm{d}r \tag{3.35}$$

Equating eqns 3.33 and 3.35, we may write the product $(n/V)g(r)$ as

$$(n/V)g(r) = p \, (D/4\pi) \, a^{-D} \, r^{D-3} \tag{3.36}$$

We cannot substitute this directly into eqn 3.31 because a real fractal is not infinite in size while the upper limit of the definite integral is at infinity and expression 3.36 has no upper bound. We need to modify 3.36 with a function that causes $(n/V)g(r)$ to drop rapidly to zero as r goes beyond some characteristic length ξ. Such a function is $\exp(-r/\xi)$. When the modified version of eqn 3.36 is substituted into eqn 3.31 we find

$$S(Q) = 1 + p\,(D/A^D)\int_0^\infty r^{D-1}\exp(-r/\xi)\frac{\sin(Qr)}{Qr}dr \qquad (3.37)$$

This integral is standard but the result is complicated and to reproduce it would not be particularly enlightening. Instead we note the values it takes in two limits.

For $Qa \gg 1$, $\qquad\qquad\qquad S(Q) \approx 1 \qquad\qquad\qquad (3.38)$

For $1/\xi \ll Q \ll 1/a$, $\qquad S(Q) \approx 1 + \text{constant}/(Qa)^D \qquad (3.39)$

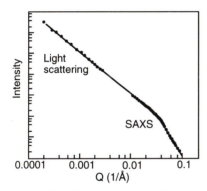

Fig. 3.7 Combined SALS and SANS data for colloidal silicate aggregates in solution.

In Fig. 3.7 we show some of the results of SALS and SAXS experiments on aggregation of SiO_2 particles of diameter $a = 2.7$ nm under acidic conditions. SALS probed the range $0.002 \le Q \le 0.02$ nm^{-1} while SAXS with $\lambda = 0.154$ nm provided a resolution such that $0.1 \le Q \le 1$ nm^{-1}. The two sets of data are combined in Fig. 3.7, scaled relative to each other so that they lie on the same line when $\log(I(Q))$ is plotted against $\log(Q)$. At small Q the gradient of this line is -2.1 and when Q is increased to about $1/a$, a cross-over to a steeper line of gradient -4.0 is seen.

Scattering from surface fractals

When the wavelength of the radiation is very short, scattering is from short-range fluctuations in the scattering density of the sample. This is largely governed by the roughness of the interface between scattering particles and the solvent. We note that in this region $S(Q) \approx 1$ (eqn 3.38) so $I(Q)$ is proportional to $F^2(Q)$. We have to be a little more careful when we evaluate $F(Q)$ with this class of material because it commonly involves porous materials in which the volume of solid matter and empty pores are comparable. In that case we can no longer assume that the radiation is scattered only by fluctuations in the density of the solid matter. Our evaluation of the fluctuations of the scattering in strength must also involve the correlation of the pores. If we substitute the integral form of $F(\boldsymbol{Q})$ (eqn 3.24) into the expression for $I(\boldsymbol{Q})$ (eqn 3.22) we find

$$I(\boldsymbol{Q}) = \frac{1}{V} \int_V \int_V \rho(\boldsymbol{r}')\, \rho(\boldsymbol{r}) \exp(\mathrm{i}\boldsymbol{Q}.(\boldsymbol{r} - \boldsymbol{r}'))\, \mathrm{d}\boldsymbol{r}\, \mathrm{d}\boldsymbol{r}' \qquad (3.40)$$

If we now write the density at an arbitrary point $\boldsymbol{r}$ as the sum of the mean density ρ_0 and a local fluctuation $\eta(\boldsymbol{r})$, and introduce the correlation function $g(\boldsymbol{r})$, we find for centrosymmetric particles

$$I(Q) = \left\langle \eta^2 \right\rangle \int_V g(r) \exp(\mathrm{i}Q.r)\, \mathrm{d}r \qquad (3.41)$$

where $\langle \eta^2 \rangle$ is the mean square of the density fluctuations. Let us now evaluate this for a porous material composed of dense matter and pores with concentrations c and $(1-c)$, and densities ρ and 0 respectively. The mean density ρ_0 is then ρ_c and the deviation of scattering density from this mean is $\rho(1-c)$ and $-\rho c$ in the solid and pores respectively. The mean squared scattering density, $\langle \eta^2 \rangle$, that appears in eqn 3.41 is then

$$\langle \eta^2 \rangle = c(\rho(1-c))^2 + (1-c)(\rho c)^2 = \rho^2 c(1-c) \qquad (3.42)$$

Remember that $\langle \eta^2 \rangle$ not only depends on the distribution of matter, it also describes the correlation of holes. We concentrate on the way in which just the matter is distributed and define a function $Z(r)$, as the probability that if the origin ($r = 0$) is in the solid material, then a point at r is also in the material. We can then write

$$Z(r) = c + (1-c)g(r) \qquad (3.43)$$

that is, when $r = 0$, $Z(r)$ is 1 and as r becomes larger, it tends to the mean concentration of matter in the sample. The concentration changes by $(1-c)$ at a rate proportional to $g(r)$. Thus, $g(r)$ may be expressed as

$$g(r) = (Z(r) - c)/(1-c) \qquad (3.44)$$

$Z(r)$ only deviates from 1 when it is within a distance r of the surface. We define a region of volume V_b within which $Z(r)$ may be smaller than 1. If V is the total volume of the sample, then the probability that it is within this boundary region is V_b/cV and the probability that it is outside, where $Z(r) = 1$, is $(cV-V_b)/cV$. We now need to calculate $Z(r)$ within the boundary region which requires an average over all the boundary region of the probability that if a point at x is within V_b so too is a point at r from this origin. We denote this probability by $p(r,x)$; if we assume that for very small r the interface looks like a plane then $p(r,x) = (r + x)/2r$ for $x \leq r$. In Fig. 3.8 we show several such cases for different x – both outside the boundary region ($x > r$), just on the boundary ($x = r$), and within the boundary region ($x < r$).

$Z(r)$ is now given by

$$Z(r) = \frac{cV - V_b}{cV} + \frac{V_b}{cV}\frac{1}{r}\int_0^r p(r,x)dx \quad (3.45)$$

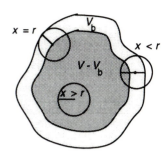

Fig. 3.8 The boundary region of thickness *r* of a particle is shown as the lighter area. The circles of radius *r* are drawn at various distances *x* from the boundary of the particle as indicated within the circles.

We can find an expression for V_b by assuming that the boundary region is built from n cubes of edge r: $V_b = nr^3$. Equation 1.21 tells how n changes with the extent of an object for a given yardstick. Conversely, for a given object, the number of particles of radius r required to fill scales with r as follows

$$n(r) = C\, r^{-D} \quad (3.46)$$

Where C depends on the size and shape of the object. Consequently we may write V_b as

$$V_b = C\, r^{3-D} \quad (3.47)$$

If we substitute for V_b in eqn 3.45 and perform the integration we find

$$Z(r) = 1 - C\, r^{3-D}/4cV \quad (3.48)$$

and, from eqn 3.44

$$g(r) = 1 - C\, r^{3-D}/4c(1-c)V \quad (3.49)$$

If we substitute this expression into 3.41 and note that $S(Q) \approx 1$, then after performing the integration we find

$$I(Q) \sim Q^{D-6} \quad (3.50)$$

Smooth particles, with $D = 2$, show a gradient of -4 when $I(Q)$ is plotted against Q. We noted this for the colloidal mass fractals built from smooth particles in Fig. 3.7. However, for a rough, porous material with surface fractal properties, such gradients are less steep and indicate that $2 \leq D \leq 3$. In Fig. 3.9 we present SAXS data for controlled pore glasses which indicate that

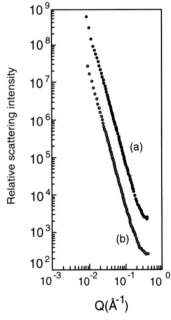

Fig. 3.9 SAXS from controlled pore glasses with pore sizes of (a) 700Å and (b) 2000Å.

$D_s = 2.20\pm0.05$ regardless of pore size over 2 orders of magnitude range in Q, and 5 in $I(Q)$.

We finish this section with a word of caution. We have assumed that the primary objects that cause the scattering are monodisperse; polydispersivity will alter the dependence of $\log(I(Q))$ on $\log(Q)$. If the polydispersivity has a power-law dependence on radius, that is if the number of elementary particles of a particular radius a scales with a as $p\ a^{-(1+D_s)}$, then we find that even when there is no fractal aggregation, $I(Q)$ depends on Q as follows

$$I(Q) \sim Q^{-(6-D_s)} \tag{3.51}$$

3.4 Energy transfer experiments

The third method that we shall consider in this chapter for the determination of D involves the transfer of energy from an excited molecule called the donor (B) to another molecule called the acceptor (A). The rate at which energy is transferred depends on the spatial distribution of A about B and hence depends on D. The most common type of experiment concerns two types of dye molecules placed on a fractal surface. A short pulse of laser light excites one type of dye molecule to a state which may relax back to the ground state either spontaneously in a unimolecular fashion, or through the transference of energy to the other type of dye molecule. Let us look more closely at the way in which D comes into the expression for energy transfer before we consider a specific example.

The most common mechanism for the transfer of energy between B and A involves the direct coupling of electric dipoles through space. We call this direct energy transfer (DET) as opposed to a multistep process in which the energy hops between B and A in a series of steps. We consider such processes in Chapter 4. The rate of exchange of energy in a DET process between a single B molecule and a single A molecule is given in terms of their separation r by the expression

$$w(r) = \alpha\ r^{-s} \tag{3.52}$$

where $s = 6$ for dipole-dipole coupling, and may take other values for other forms of multipole coupling. If we assume initially that energy transfer between B and A is the only way in which B may relax, then the probability that B survives in the excited state for a time t after excitation, $\Phi(t,r_i)$, is given by

$$\Phi(t, r_i) = \exp(-t\ w(r_i)) \tag{3.53}$$

where r_i is the separation between B and an atom of A, located at the site i. If the probability that site i is occupied is p then $\Phi(t, r_i)$ becomes the probability that i is not occupied by A plus the probability that transfer does not occur in time t, given that the site *is* occupied i.e.

$$\Phi(t,r_i) = (1-p) + p\ \exp(-t\ w(r_i)) \tag{3.54}$$

If we sum over all the available sites i and assume their occupancy to be indendent of each other, the total survival probability, $\Phi(t,r)$, is given by the product of terms like 3.54 i.e.

$$\Phi(t,r) = \prod_i \{(1-p) + p \exp(-tw(r_i))\} \qquad (3.55)$$

If p is very small we may simplify eqn 3.55 through the approximation

$$\prod_i (1-\delta_i) \sim \prod_i \exp(-\delta_i) \sim \exp(-\sum_i \delta_i) \qquad (3.56)$$

where $\delta_i \ll 1$. Equation 3.56 may then be rewritten

$$\Phi(t,r) = \exp(-\sum_i (1 - \exp(tw(r_i)))) \qquad (3.57)$$

If we now assume that the mass of the object on which B and A are placed is distributed continuously with a fractal distribution D, and that B and A coat the object homogeneously, we may use eqn 1.22 and replace the sum of eqn 3.57 with the following integral

$$\Phi(t,r) = \exp(-p' \int (1 - \exp(tw(r))) r^{D-E} dr) \qquad (3.58)$$

If we substitute for $w(r)$ from eqn 3.52 and perform the integration above, we find

$$\Phi(t,R) = \exp(-pAt^{D/s}) \qquad (3.59)$$

where A is a constant that is independent of t. Finally, if we allow the possibility that B may spontaneously decay at a rate w_0 without energy transfer then eqn 3.59 is modified to become

$$\Phi(t,R) = \exp(-(tw_0 - pAt^{D/s})) \qquad (3.60)$$

There are several conditions that must be met for this expression to be valid and useful. First, the concentrations of B and A must be such that [B]<<[A]<<1 and we must ensure that their distribution reflects the distribution of mass in the fractal host or substrate faithfully. The range over which transfer is efficient compared with spontaneous decay should be as large as possible to extend the range of r over which D is probed. We must be able to determine w_0 for B unambiguously and there should not be significant luminescence from the A molecules in the part of the spectrum in which fluorescence from B is monitored

One system that has been thoroughly studied in this manner comprises the dye molecules rhodamine 6G (R6G) and malachite green (MG) as B and A respectively, adsorbed on a series of porous silica glasses. We reproduce the fluorescence spectrum of R6G and the adsorption spectrum of MG in Fig.

3.10, showing the overlap that permits transfer to occur. The quantum yield for fluorescence from R6G is very high compared with that of MG, so any fluorescence may be attributed to R6G. The porous glasses – silica 40, silica 60, silica 100, and silica 500 – contain a distribution of pore diameters whose centre is given in Ångstroms by the number in their name.

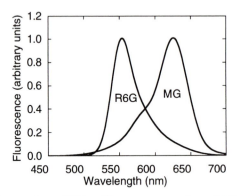

Fig. 3.10 Absorption spectrum of MG and fluorescence spectrum of R6G adsorbed on silica 100.

The glass was first coated with R6G or a mixture of MG and R6G by standing it for 24 h in acetonitrile solutions of the dyes, followed by drying. The concentration of the dyes was determined spectrophotometrically. The samples were excited electronically with an argon ion laser which delivers a continuous train of 10^{-10} s pulses of light with a wavelength of 514 nm. Fluorescence from R6G was monitored at 560 nm and its decay determined for samples with and without MG (Fig. 3.11).

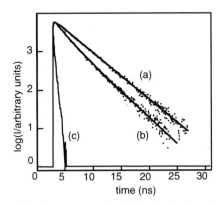

Fig. 3.11 The decay of the fluorescence from R6G adsorbed on silica 100 (a) with MG present and (b) without MG present. The dots represent the experimental data and the lines least-squares fits to theoretical decay curves for the determination of D. The exciting pulse is depicted by the sharp spike (c) that decays before 5 ns.

Expression 3.60 was fitted to the two sets of decay data for each type of glass to produce estimates of w_0 and D, assuming that $s = 6$. The range over which transfer is efficient is believed to be about 8 nm. This is smaller than the average pore diameter for silica 100 and silica 500 so we might expect that DET measurements would probe the surface of the silica particles from which the solid is constructed. Measurements indicate D to be close to 2 for these glasses, which is consistent with the belief that the silica particles are smooth when studied with the resolution of the DET measurements. A higher value of D was found for the silica 40 sample whose pores are sufficiently small for the DET to be sensitive to the fractal porous structure.

3.5 Summary

Box-counting and walking techniques may be used to determine fractal dimensions for objects which may be photographed and which have $D \leq 2$

Small angle scattering techniques may be used to estimate D for both surface and mass fractals. When the scattering vector Q falls in the range $QR \gg 1$ and $Qa \ll 1$, where R and a are the radius of the clusters and elementary particles in monodisperse aggregates in solution, $I(Q)$ provides information about the distribution of mass within the cluster: a graph of $\log(I(Q))$ against $\log(Q)$ gives a gradient of $-D$.

When the scattering vector is such that $Qa \gg 1$, $I(Q)$ is sensitive to the surface between the solid and its environment and a plot of $\log(I(Q))$ against $\log(Q)$ gives a straight line of gradient $D-6$. The range of length-scales for which such methods are most useful ranges from 0.5 – 500 nm

Measurements of the rate of direct energy transfer between dye molecules adsorbed on a surface may provide an estimate for the fractal dimension of that surface for length-scales up to about 10 nm.

Further reading

In addition to the book edited by Avnir and that by Kaye to which I referred in Chapter 1, you may find the following references useful.

Guinier, A. and G. Fournet, G. (trans. Walker C.B. and Yudowitch, K.L.) (1955). *Small-angle scattering of X-Rays*. Wiley, New York.
Feigin, L.A. and Svergun, D.I. (1987). *Structure analysis by small-angle X-Ray and neutron scattering*. Plenum, New York.
Martin, J.E. and Hurd, A.J. (1987). Scattering from fractals. *Journal of Applied Crystallography* **20** 61.
Schaeffer, D.W. (1987). Small-angle scattering from disordered systems. *Materials Research Society Symposia Proceedings* **79** 4.
Blumen, A., Klafter, J., and Zumofen, G. (1986). Models for reaction dynamics in glasses. *Optical spectroscopy of glasses*. (Zschokke, I. (ed.)). D. Riedel, Dordrecht.

Problems

3.1. The growth of fumed silica by burning $SiCl_4$ rapidly in H_2 and O_2 was introduced in Chapter 2. The first stage of growth involves the formation of silicic acid, $Si(OH)_4$, which then polymerises to produce small, rough primary particles. As these particles move further from the nozzle of the flame, the density of aggregating particles decreaes, and further growth probably involves DLA of these particles to produce larger, low-density clusters. The growth conditions may be varied to produce silica samples with different types of primary particle, and different densities of primary particles. SANS data for three different grades of the fumed silica Cab-O-Sil, M-5, HS-5 and EH-5 are presented in Fig. 3.12. The areas per unit mass of different grades were estimated by a gas coverage technique as 200, 325 and 390 m^2g^{-1} in no particular order. What is the surface and mass fractal dimension of the three forms of silica, taking care to specify the appropriate length-scales. Comment on the likely size of the primary clusters and of the overall size of the particles. Specify which surface areas apply to which samples.

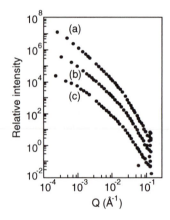

Fig. 3.12 SAXS data for problem 3.1; curves (a), (b), and (c) are for different fumed silicas.

3.2. The polymerisation of tetraethoxysilane (TEOS) in water and ethanol to produce silica gels is very sensitive to pH and the amount of water present. In basic solution, the first step is the conversion of Si—OEt to Si—OH by nuclophilic substitution which involves an inversion of the geometry at the Si centre. The Si—OH functional site may then form Si—O—Si' links by a condensation reaction with Si'—OEt. This disfavours further hydrolysis of Si—OEt because inversion is hindered. Thus, the degree of functionalisation, and therefore the degree to which the polymer branches, depends on the degree to which $Si(OEt)_4$ hydrolyses.

Fig. 3.13 shows the results of SAXS experiments performed on silica solutions prepared from TEOS in basic ethanol solution with the different H_2O:TEOS ratios 1, 2, 3, and 4 as well as a sample of stabilised colloidal silica particles. Assign the different curves to the different samples, and discuss the results in terms of the likely growth processes.

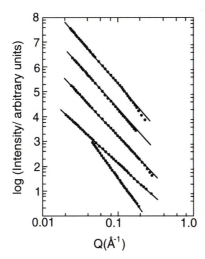

Fig. 3.13 SAXS data for colloidal silica and for four different samples of silica gels prepared through the hydrolysis of TEOS with different amounts of water present.

3.3. Molecules of the dye Rhodamine 6G (R6G) are adsorbed on a sample of the porous glass Vycor and stimulated with a laser pulse to produce fluorescences. The decay of the fluorescence with time t is proportional to $\exp(-\alpha t)$, where α is determined by a least-squares fit to be $(2.75\pm0.02) \times 10^8 \text{s}^{-1}$. When the glass sample is also exposed to a small amount of the dye Malachite Green (MG), which has an optical absorption band that overlaps with the fluorescence band of R6G, the fluorescence life-time of the R6G is greatly reduced and its survival probability no longer has a simple exponential form but decays as $\exp(-\alpha t + \beta(\alpha t)^\gamma)$. The value of γ changes with the concentration of MG, ranging from 0.45 at low concentrations to 0.40 at higher concentrations.

What is the radiative lifetime of R6G in samples free from MG, and what do these data indicate is the fractal dimension for the surface of Vycor at (a) low and (b) high concentrations of MG? Comment on the differences between (a) and (b).

4 Chemistry in fractal environments

4.1 What use are fractal concepts?

In the previous chapters I have tried to illustrate what a fractal is, and how it may be grown and characterized. Along the way I have made implicit and explicit statements about the way in which the concept of a fractal dimension may be used. Its principal function is to provide a language for the description of irregular objects and an index that gives a rough-and-ready measure of the degree to which such objects deviate from some Euclidean ideal as a function of the length-scale. In this chapter we look more closely at such applications and show how the fractal dimension may also help us to describe physical and chemical processes in and on fractal objects. I have been selective in my choice of applications, restricting the discussion to subjects which are closest to the experiences and needs of most chemists. Even within this subset of applications I have had to be selective. Explicit descriptions of the fractal properties of polymers, electrode surfaces, and geochemical objects will not be given here, and the reader is left to apply the lessons of the previous chapters to these subjects when the need arises.

Perhaps the most significant application of fractal concepts which I will neglect concerns temporal rather than spatial fractals, that is fractals which have a dilational symmetry in time rather than space. We touched on the subject in Chapter 1 when the Cantor dust was related to the frequency of noise in an electrical transmission line. The application of fractal concepts to noise, turbulence, and chaotic behaviour is treated in some of the references at the end of the chapter.

4.2 Adsorption of fluids on fractal surfaces

The concept of a surface fractal was introduced in Chapter 1 by showing how the effective area of many porous objects measured by adsorption of gas molecules depended on the size of the molecules. This type of process may not only be used to estimate the fractal dimension of a surface, but may also have a direct bearing on the rate of many heterogeneous reactions. In this section we consider how the effective area of a surface scales with the size of the probe molecule, and also how it depends on the strength of the interaction between the fluid molecules and the surface. Weak Van der Waals interactions give rise to physisorption in which the fluid covers all of the surface that is physically accessible, while the stronger covalent interactions responsible for chemisorption only involve a subset of all the available surface sites. We consider both processes in the following sections, and then go on to see how the geometry of an interface may influence the static aspects of chemical reactions at a surface. Dynamic aspects will be considered in Sections 4.4, 4.5 and 4.6.

Physisorption

Physisorption is most commonly studied by measuring the volume *V* of a gas that is adsorbed on a surface as a function of its pressure *P* at a constant temperature. The relationship between *V* and *P* is called the adsorption isotherm. Surface scientists often try to use such information to reveal what interaction there is between the surface and the adsorbate at an atomic level. There are two widely used expressions which attempt to provide a link between experiment and molecular interactions. The first and simplest is due to Langmuir and is concerned with the case in which only a monolayer may be adsorbed. It is assumed that the energy of adsorption is the same for every site and independent of the fractional coverage θ. If it is also assumed that the adsorbed molecules are in equilibrium with those in the gas phase then θ is given by:

$$\theta = KP/(1+KP) \tag{4.1}$$

where *K* is the ratio of the adsorption rate constant k_a to the desorption rate constant k_d.

The second common expression is due to Brunauer, Emmett and Teller and is called the BET isotherm. I will not give a derivation here and leave the interested reader to look for it among the reading material given at the end of the chapter. However, it is worthwhile considering the basis and form of the expression, and see how and why it is influenced by the fractal character of a surface. The major difference between the foundations of the BET isotherm and the Langmuir isotherm lies in the consideration of multilayer adsorption for the former. The rate constants for adsorption and desorption of the second and any subsequent layers take the same values k_a' and k_d' respectively. At any point on the isotherm it is assumed that adsorption and desorption are in equilibrium for every layer. The ratio of the total number of adsorbed gas molecules, *n*, to the number required for monolayer coverage, n_{mono}, is then given by

$$\frac{n}{n_{mono}} = \frac{c\sum_{i=1}^{\infty} ix^i}{1+c\sum_{i=1}^{\infty} x^i} \tag{4.2}$$

where $x = k_a'/k_d'$ and $cx = k_a/k_d$. *x* is also equal to P/P_0, the ratio of the pressure of the gas to the saturated vapour pressure of pure liquid adsorbate at the same temperature. The sums in eqn 4.2 may be evaluated and the equation rearranged to produce

$$P/n(P-P_0) = 1/(n_{mono}c) + (c-1)/(n_{mono}c)(P/P_0) \tag{4.3}$$

If the left-hand side of eqn 4.3 is plotted against P/P_0 then at smaller values of P/P_0 a straight line may be found whose gradient and intercept provide estimates for c and n_{mono}. If the surface area σ of the molecule is known, the area of the surface may then be determined.

If the surface has fractal scaling properties then the BET model must be modified. Consider Fig. 4.1 which depicts the effect of covering a fractal surface with successive molecular coats. As the thickness of the coverage increases, so the finer details of the surface are lost. An analogy that is useful, if not strictly accurate, is the smoothing of the features of a landscape as snow settles. Successive layers may also have different volumes, and this will affect the number of adsorbate molecules that may be accommodated as deposition progresses.

Suppose the surface is covered in a uniform coat of spheres of radius ρ such that ρ lies within the range of length-scales for which fractal scaling applies. The number of molecules lying on a surface of linear extent λ is proportional to $(\lambda/\rho)^D$, so the volume of the coat, $V(\rho)$, is given by

$$V(\rho) \sim (\lambda/\rho)^D \, \rho^3 = p \, (\lambda/\rho)^D \rho^3 = p \, \lambda^D \, \rho^{3-D} \qquad (4.4)$$

where p relates to the way in which the spheres pack on the surface (eqns 1.21 and 1.22). The effective area $A(\rho)$ covered by the coat is given by

$$A(\rho) = \delta V(\rho)/\delta\rho = p \, (3{-}D) \, \lambda^D \rho^{2-D} \qquad (4.5)$$

In the BET model the surface may be covered in several layers of molecules so we may substitute the spheres of radius ρ by i layers of smaller spheres of radius r which represent the molecules. We then set $\delta\rho$ in eqn 4.5 to r, and ρ to ir. The ith layer has an effective volume $\delta V(r)$ given by

$$\delta V(r) = p \, (3{-}D) \, \lambda^D \, i^{2-D} r^{3-D} \qquad (4.6)$$

The number of molecules in the ith layer, $n_i(r)$, is proportional to $\delta V(r)$ divided by the volume of the probe molecules:

$$n_i(r) = p' \, (3{-}D) \, \lambda^D \, i^{2-D} \, r^{3-D} \qquad (4.7)$$

The number adsorbed in the first coat, $n_1(r)$, is given by substituting 1 for i in eqn 4.7:

$$n_1(r) = p' \, (3{-}D) \, \lambda^D \, r^{3-D} \qquad (4.8)$$

The way in which $n_1(r)$ changes with i is then described as follows

$$n_i(r)/n_1(r) = i^{2-D} \qquad (4.9)$$

Fig. 4.1 Illustration of the effect of covering a bare fractal surface with successive layers of molecules, represented by disks.

Therefore, as i increases for a surface with $D > 2$, the number of molecules that may be adsorbed per layer decreases. The expression for the BET isotherm (eqn 4.2) is then modified as follows:

$$\frac{n}{n_{\text{mono}}} = \frac{c \sum_{i=1}^{m} i^{2-D} \sum_{j=1}^{m} x^j}{1 + c \sum_{i=1}^{m} x^i} \qquad (4.10)$$

Note that we have also replaced the upper limit of the summations by a finite number m.

The Langmuir isotherm only involves monolayer adsorption so the correction we made to the BET isotherm for the reduction in available volume per layer is not relevant.

The appearance of the BET isotherm changes with D in the manner illustrated in Fig. 4.2(a). At low coverages, the rough fractal and the smooth surfaces show very similar behaviour, but as the proportion of sites with multilayer coverage increases, it becomes progressively harder to deposit more molecules as D is increased. The graphical technique described under eqn 4.3 becomes much less satisfactory as D rises, and the linear region may be too poorly defined to determine c and n_{mono} properly. In such circumstances it is better to fit the entire isotherm to eqn 4.10.

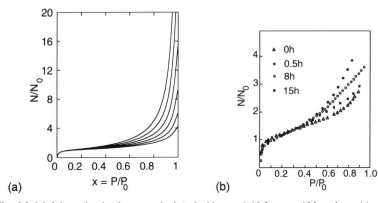

(a)   (b)

Fig. 4.2 (a) Adsorption isotherms calculated with eqn 4.10 for $m = 100$ and c set to 100 as for the different values of the fractal dimension D marked by each curve. (b) Adsorption isotherm measured for a pillared clay as the layers are separated by increasing its exposure to acidic solution. As the time increases, the dimension of the solid falls.

In Fig. 4.2(b) we show the N_2 BET isotherms for a form of clay in which silicate sheets are held together with pillars of a metal oxide such as Fe_2O_3. When 1M H_2SO_4 is introduced, the pillars are slowly destroyed, and the material is transformed from a porous solid with D close to 3, to a collection of separated sheets with D close to 2. Further values of D obtained through this type of gas adsorption measurements for a variety of surfaces is given in Table 4.1.

The interpretation of gas coverage data contains several pitfalls which the reader may already have spotted.

Table 4.1 Fractal dimension of selected surfaces determined by adsorption measurements using probe molecules with a range of surface areas. All the materials are described in this chapter and Chapter 1 except for graphon which is a particularly smooth form of graphite, and faujasite which has a similar framework to zeolite Y. When the probe molecule is an alkane, its size is given in terms of the number of carbon atoms in the backbone.

Surface material	Probe molecule	Fractal dimension
silica 60	n-alkanes ($C_1 - C_6$)	2.98±0.24
silica 60	'spherical' alcohols	2.97±0.02
silica 100	n-alkanes ($C_1 - C_4$)	2.90±0.40
cpg 75	n-alkanes ($C_1 - C_7$)	2.09±0.08
graphon	N_2 and n-alkanes (C_{22}, C_{28}, C_{32})	2.07±0.01
synthetic faujasite	N_2 and n-alkanes ($C_3 - C_7$)	2.02±0.05

First we must already know what the effective surface area σ of the probe molecule is. For a simple molecule such as Ar, an estimate may reliably be made from measurements of the density of the liquid; this may not be the case for larger molecules. If the molecule is particularly anisotropic in terms of the cross-sectional area it presents for different aspects, we need to know its orientation relative to the surface. Furthermore, a large molecule may adopt several conformations with quite different surface areas. In general we turn the problem on its head, and use a well-characterized surface in conjunction with adsorption data for the poorly understood molecule to elucidate its properties. Thus, if the gradient and intercept in Fig. 1.8 are determined through some well-characterized molecule, then measurement of n for the unknown materials provides an estimate for the effective value of σ. A good example of this procedure is provided by measurements of the adsorption of n-alkanoic acid molecules of the form $C_nH_{2n+1}CO_2H$ on silica. The surface was characterized by covering it with samples of different highly branched alkanols whose shape was taken to be spherical. The cross-section of the alkanoic acid molecules was estimated from Van der Waals radii so that the empirically determined value for the effective surface area could be used to reveal the orientation adopted by the acid molecules. It was found that the chain-like molecules lay with the chain axis *parallel* to the silica surface. It is believed that this orientation is due to a more favourable H-bonding interaction between the bidentate carboxyl group and either geminal or vicinal OH groups bound to silicon atoms on the surface (Fig. 4.3). Similar measurements may be used to determine the conformation of polymers adsorbed on surfaces.

The range of length-scales over which D may be probed is limited to the range of suitable probe molecules which, for reasons given in the previous section, does not extend much beyond cross-sectional areas of $50\,\text{Å}^2$. Extrapolation below the radius of the smallest probe particles is also not good practice. Above a critical value of the diameter of the probe molecule, many porous solids show a sharp reduction in the surface available owing to bottlenecks in the form of narrow entries to the interior of the solid. By comparison, non-intrusive methods such as SAXS and SANS measurements show much higher surface area at comparable length-scales.

So far, we have considered scaling the size of the probe molecule for a given surface. Alternatively, we may fix the probe molecule and expose it to porous particles with various size distributions. We noted in Section 1.3 that if we took a sample of mass M in the form of a powder comprising particles whose area scaled with radius as r^D, then the ratio of the total area A to M is proportional to r^{D-3} (Eqn 1.28). Such measurements generally scan a wider range of length-scales than is feasible in methods in which the size of the probe molecules is changed.

Chemisorption

In a physisorption process, molecules probe all the surface that is accessible. Chemisorption of molecules of a comparable size only uses the fraction of this surface that can form chemical bonds with the molecule. The most important case for a chemist concerns the activity of fine metal particles used in heterogeneous catalysis. It is usual for particles to be supported on a porous material, but the particles themselves are generally fairly smooth and their surfaces have a fractal dimension close to 2.0 over the range of length-scales of common probe molecules. Examples of the fractal dimension measured for a selection of metal particles, supports, and gas molecules is given in Table 4.2. It is remarkable that the values are not only smaller than might be expected for physisorption, but are mainly less than 2.0. This is believed to be because of the way in which the particular sites which are active towards chemisorption scale with the particle radius r. Chemisorption bonds are often formed at edges or corners on the metal surfaces and the number of these sites will scale with r as r^0 and r^1 respectively. This will reduce the measured value of D below 2.0. It also means that the value of D for a particular surface depends on the nature of the adsorbed molecule because another molecule may have a different selectivity towards different surface sites, even if it has an identical radius. In such circumstances D may be called the fractal dimension for chemisorption, but rather than define a new symbol we remember that the empirical quantity may have a slightly different interpretation from the one we gave D when we discussed physisorption.

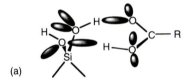

(a)

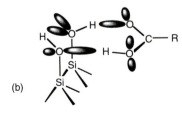

(b)

Fig. 4.3 Orientation of alkanoic acid molecules on the surface of silicates. The flat disposition of the molecule leads to more favourable H-bonding between the bidentate carboxylic acid functional group and either (a) geminal or (b) vicinal –OH groups on the surface.

Table 4.2 Scaling dimension for chemisorption for a variety of dispersed metal catalysts. Silica Z is a non-porous form of silica.

Metal-support and gas		Range of particle size (nm)	Fractal dimension D
Pt-SiO$_2$	H$_2$	1.3–4.0	1.67±0.05
Pt-SiO$_2$	CO	6.6–27	1.6±0.2
Pt-Al$_2$O$_3$	H$_2$	1.3–10	1.91±0.03
Ag-Cab-O-Sil	O$_2$	6.0–54	2.06±0.06
Ag-silica Z	O$_2$	5.5–40	1.82±0.07

4.3 Reactions at surfaces

Just as there is no clear division between physisorption and chemisorption, so the boundary between chemisorption and chemical reactions at surfaces is indistinct. The unimolecular decomposition of a diatomic molecule adsorbed on a metal surface may be viewed as an extreme case of chemisorption in which the bonds formed between the molecule and the surface weaken the intramolecular bond. If we restrict the discussion to reactions in which the surface acts as a heterogeneous catalyst, and is not changed chemically in a permanent way, then we may break the reaction down into five basic steps.

1.   Diffusion of reactants to the surface.

2.   Adsorption of reactant or reactants on the surface.

3.   Reaction of or between reactants.

4.   Desorption of reaction products.

5.   Diffusion of products from the surface.

The rate-determining step is usually 3 and there are commonly two forms of such a reaction. First, the reaction may exclusively involve adsorbed species. This is called a Langmuir–Hinshelwood (LH) reaction. If we consider the following bimolecular example of a LH reaction

$$A + B \longrightarrow P \tag{4.11}$$

we usually describe the rate using the following equation

$$d[P]/dT = k\theta_A\theta_B \tag{4.12}$$

where θ_A and θ_B are the fractional coverages of A and B. Alternatively, step 3 may involve reaction between an adsorbed species A and a gas molecule B that collides with that region of the surface. This is an example of the Eley–Rideal (ER) mechanism. The rate law for an ER process is

$$[P]/dT = k\theta_A P_B \tag{4.13}$$

Both expressions (4.12 and 4.13) depend on the extent of surface coverage so we might expect the activity of a heterogeneous catalyst to be partly governed by principles similar to those for chemisorption. The activity of powdered catalyst per unit mass scales with particle radius r according to eqn 1.28:

$$Activity \sim r^{D_R - 3} \tag{4.14}$$

where D_R is called the reaction dimension. In Table 4.3 we show values for D_R for a variety of metals supported on porous materials and catalysing reactions such as the reduction of alkanes or alkenes with hydrogen. It is remarkable that D_R spans a wide range of values, from close to zero to almost 6.

Table 4.3 Reaction dimension for a variety of reactions catalysed by dispersed metal particles.

Reaction	Catalyst and support	Reaction dimension D_R
Benzene hydrogenation	$Pt–Al_2O_3$	2.05 ± 0.02
Propene hydrogenation	$Pt–SiO_2$	1.58 ± 0.03
Cyclopropane hydrogenolysis	$Pt–Al_2O_3$	2.29 ± 0.07
Cyclopropane hydrogenolysis	$Pt–SiO_2$	1.85 ± 0.06
Oxidation of ethene to ethene oxide	Ag–Cab-O-Sil	1.16 ± 0.11
Oxidation of ethene to CO_2	Ag–silica Z	0.71 ± 0.16
Reduction of CO with H_2 to CH_4	Pd–Cab-O-Sil	$2.9\pm.015$
Reduction of N_2 to NH_3	Fe–MgO	5.8 ± 0.5

In all these cases, it is believed that the accessible surface area for physisorption of the reactants has a dimension close to 2.0. Suppose now that catalysis could only occur at a crystallite vertex, edge, or plane. As we indicated for chemisorption, the number of such sites scales with r as r^0, r^1 or r^2 respectively so that as r grows the increase in the number of active sites is smaller than or equal to the growth in the total number of surface sites.

Low, non-integral values for D have been interpreted as reflecting a mixture of active sites, and may even be used to predict just what proportion of each is involved in the reaction. Thus, for the case of the oxidation of ethene with O_2 on silver dispersed on a non-porous form of silica called silica Z and on Cab-O-Sil, D_R was found to be 1.58 ± 11 and 1.16 ± 11, respectively, when the product was ethene oxide, and 0.71 ± 16 and 0.43 ± 18, respectively, when the product was CO_2. The individual particles of silver grow in regular forms such that the free energy is minimized. The most favourable shape is the cubo-octahedron depicted in Fig. 4.4. The near–spherical form ensures that the greatest number of metal–metal bonds form in the bulk of the cluster, while the surface is formed from the (100) and the (111) planes which are the most highly coordinated. The relation between the number of face, edge, and vertex sites as a function of particle size may be calculated precisely, and our estimates of the relative activity of each of these types of sites may be varied so as to reproduce the observed form of the scaling of active area with r.

What interpretation may then be given to reaction dimensions *greater* than D? Clearly, the process involved relies on sites whose populations increase with r *faster* than the surface area. A specific example is the reduction of CO to methane with hydrogen gas over palladium supported on silica in the form of Cab-O-Sil. Once more, the crystallites of palladium adopt a cubo-

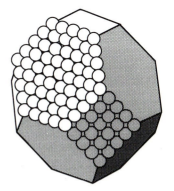

Fig. 4.4 Cubo-octahedron shape adopted by small metal particles. The atoms that form in (100) and (111) planes have been drawn as spheres.

octahedral form for which the relative reactivities of the (111) and (100) faces adopt the ratio 2.8:1. The ratio of the number of atoms on these sites increases with r, so that as the average particle size increases the surface reactivity per unit mass of catalyst increases as indicated in Fig. 4.5.

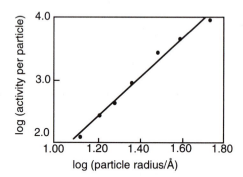

Fig. 4.5 Determination of the reaction dimension for the reduction of CO to CH_4 by H_2 over Pd supported on Cab-O-Sil. The units of the activity are proportional to mol s^{-1} particle^{-1}.

The LH and ER rate laws (eqns 4.12 and 4.13) depend not only on surface coverage, but also on a rate constant. Your experience of reaction kinetics will probably be centred on homogenous reactions in which the rate constant is related to the rate at which reactants pass over an activation barrier on the free energy surface to become product molecules. However, this process may be very fast compared to the rate with which the reactive atoms or molecules diffuse together. Such reactions are said to be *diffusion-controlled* as opposed to *activation-controlled*; we might anticipate that the kinetics of a diffusion-controlled reaction will be described by means of the diffusion equations we derived in Chapter 2 for particles executing random walks. We will now consider how a random walk is influenced by the fractal dimension of the environment.

4.4 Diffusion in fractal environments

The rate of a diffusion-controlled reaction may depend on the concentration of the reactants and the number of different sites that the reactants visit in a time t. This tells us how probable it is that the reactants collide within that time and is given the symbol $S(t)$. The rate constant for the reaction is proportional to the rate at which the reacting species visit these new sites, i.e.

$$k \sim \mathrm{d}\,S(t)/\mathrm{d}t \qquad (4.15)$$

The mean-squared displacement for a random walk in any direction was given in eqn 2.20, and implies that for isotropic migration in E dimensions, we may define a mean volume $V(t)$ explored in time t as follows:

$$V(t) \sim (\Gamma t)^{E/2} \qquad (4.16)$$

If the characteristic spacing between sites is a, then $V(t)$ contains of the order of $V(t)/a^E$ different sites. The number of jumps taken in that time, $N(t)$, is linearly proportional to time so we may write

$$N(t) \sim t \tag{4.17}$$

The time spent per site is given by the ratio

$$N(t)/N_s(t) \sim a^E t/V(t) \sim t^{1-E/2} \tag{4.18}$$

Where $N_s(t)$ is the number of different sites visited in a time t. This tells us that for $E < 2$, the time spent at any site increases with time: such a random walk is called 'compact' or 'recurrent'. Alternatively, for $E > 2$, the time spent at any site decreases with time and the random walk is called 'non-compact' or 'non-recurrent'. A different way of expressing this is to say that for $E < 2$, the probability that a migrating particle will return to a particular site tends to 1 as t tends to infinity, but for $E > 2$ it tends to zero. A common analogy used in the field of random walks concerns a drunk ejected from a bar. If the bar is in a narrow alley which constrains the walk to one-dimension, the drunk does not wander very far and will certainly return at a future time. However, if the drunk could fly, and flutter randomly in three dimensions, the probability that he or she eventually finds their way back to the bar tends to zero as t tends to infinity.

The statistical properties of random walks on lattices of different Euclidean dimension may be analysed to determine $S(t)$. The results are:

$$S(t) \sim t^{1/2} \qquad \text{for } E = 1 \tag{4.19}$$

$$S(t) \sim t/\ln t \qquad \text{for } E = 2 \tag{4.20}$$

$$S(t) \sim t \qquad \text{for } E = 3 \tag{4.21}$$

It was once believed that eqn 4.19 could be extended to fractal dimensions less than 2 as follows:

$$S(t) \sim t^{D/2} \qquad \text{for } D < 2 \tag{4.22}$$

However, in order to make this leap, it is necessary to assume that a particle diffusing through or over a fractal object is able to step uniformly between sites. When we remove translational symmetry from the space in which the walk occurs, the ability to step in a given direction may not have a uniform probability. In order to appreciate how this may influence $S(t)$ let us consider the following regular (Fig. 4.6(a)) and disrupted (Fig. 4.6(b)) square lattices.

The regular lattice with $E = 2$ has a translational symmetry and Γ is independent of r. If we remove bonds from Fig. 4.6(a) at random such that the concentration of bonds is p, then there is a critical concentration, p_c, below which there is no continuous pathway provided by the bonds from one side to the other. Imagine that the bonds are made of metal wire and that the links are cut randomly. When $p < p_c$, the object will no longer conduct electricity from one side to the other (Fig. 4.6(c)). We call p_c the percolation threshold (the word *percolate* is derived from the Latin word *percolare* which means 'to strain through'). Clearly, for this object, p_c depends on the way in which we chose to cut the bonds. Fig. 4.6(d) shows an object with $p = p_c$ which is clearly divided into two portions. However, this choice of severed bonds has a very low probability and the analogous case for a large object with of the order of 10^{23} bonds is so improbable that we discount it.

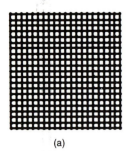 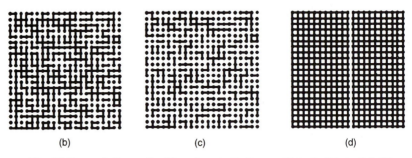

(a) (b) (c) (d)

Fig. 4.6 Square lattices with different concentrations p of bonds: (a) $p = 1.0$; (b) $p = p_c$; (c) $p < p_c$; (d) $p > p_c$ but with a very improbable choice of cut bonds.

For a large object, the value of p_c takes a precise value that depends only on the type of lattice and the way in which it is disrupted. The removal of bonds is less disruptive than the removal of a site and p_c is accordingly always higher for site dilution than bond dilution. Values of p_c are given for several common lattices in Table 4.4. If we return to the analogy of the drunken walker, migration on a fractal corresponds to placing the bar in a particularly tortuous maze. The drunk explores dead ends and closed loops as well as through routes, and this impedes progress from the starting point.

Table 4.4 Percolation thresholds for bond and site dilution on selected lattices. Values are given to two decimal places except for those that take the exact value 1/2

Lattice	Dimension	Site	Bond
Square	2	0.59	0.5
Triangular	2	0.5	0.35
Diamond	3	0.43	0.39
Primitive cubic	3	0.31	0.25
Body-centred cubic	3	0.25	0.18
Face-centred cubic	3	0.20	0.12

What relevance does this have to our problems with fractals? The object in Fig. 4.6(b) contains dense regions and holes, which appear to have some statistical dilational symmetry. It has been shown that an object at the percolation threshold is a fractal of dimension $D = 91/48 \approx 1.896$ and $D \approx 2.5$ when the parent lattice has $E = 2$ or 3 respectively. The percolating object provides a convenient model with which to explore diffusion processes on objects with a known value of D. The same could be said of many of the deterministic fractals we introduced in Chapter 1. However, the percolating object has the advantage that it is relatively easy to mimic with real materials and then test our understanding of diffusion on fractal and disordered materials. One way of realizing such a fractal is to crystallize a mixture of two isomorphous materials A and B for which mole fraction of A is equal to p_c. If the site occupancy in the resultant solid is random then the sublattice of A will be a fractal. The backbone of a polymer at its gelation point is also believed to be a percolating cluster.

If we recall the diffusion equation (2.20) and extend it to E dimensions by noting that $<r^2(t)> = <x^2(t)> + <y^2(t)> + <z^2(t)>$, then we may write the diffusion constant Γ as proportional to the rate at which $<r^2(t)>$ changes with time, i.e.

$$\Gamma \sim d/dt <r^2> \tag{4.23}$$

However, for a fractal object, we have observed that Γ gets smaller as r increases. It is common to express this diminished mobility as

$$\Gamma \sim r^{-\theta} \tag{4.24}$$

where $\theta > 0$. The classical diffusion case corresponds to $\theta = 0$. If we substitute eqn 4.24 in eqn 4.23, rearrange the result, and integrate, we find

$$<r^2> \sim t^{(2/(2+\theta))} \tag{4.25}$$

The volume swept out in a time t now becomes

$$V(t) \sim <r^2>^{D/2} \tag{4.26}$$

For a recurrent walk, $S(t)$ and $V(t)$ are equivalent, so if we combine eqns 4.25 and 4.26 we find

$$S(t) \sim t^{D/(2+\theta)} \tag{4.27}$$

This equation is commonly rewritten in terms of a new quantity called the *fracton dimension D_s* as follows

$$S(t) \sim t^{D_s/2} \tag{4.28}$$

where

$$D_s \equiv 2D/(2+\theta) \tag{4.29}$$

D_s characterises the diffusive character of the random walk on a fractal object of dimension D. Recall that for a Euclidean object, $D = E$ and $\theta = 0$ so that $D_s = E$. In general, $\theta > 0$ and the following inequality holds

$$D_s \leq D \leq E \qquad (4.30)$$

The precise value of θ and hence D_s clearly depends on the distribution of diffusion pathways averaged over the lattice and we might expect this to depend on the particular network. In many problems of interest, diffusion occurs on a percolating network for which it has been shown that D_s then adopts a universal value of 4/3 regardless of lattice type for embedding dimensions $E = 2$ and 3.

The criterion for the walk to be recurrent is that $D_s < 2$. If $D_s > 2$, we cannot define a volume that has been swept out and $S(t)$ grows linearly with t as described in eqn 4.21.

4.5 Diffusion-controlled reactions in fractal environments

We are now in a position to describe the rate of diffusion-controlled reactions in fractal environments. We will first consider ER reactions in which particles diffuse through Euclidean space until they collide and react with molecules on a fractal surface. We then look at bimolecular homonuclear and heteronuclear LH reactions in fractal and non-fractal environments.

Eley–Rideal reactions at fractal surfaces

We consider an ER reaction in which B molecules perform a random walk in $E = 3$ dimensional space until they hit the fractal surface of dimension D where they may react with adsorbed molecules A to produce product molecules P (eqn 4.11). In a time t B travels a mean distance r given by eqn 4.25 with $\theta = 0$. This allows us to define a volume $V(r)$ which includes the region that lies a distance r from the surface, and which is therefore proportional to the number of molecules that may diffuse to the surface in a time $t^{1/2}$. Equation 4.4 tells us that $V(r)$ scales with r as follows:

$$V(r) \sim r^{3-D} \qquad (4.31)$$

We may rewrite this in terms of time as

$$V(t) \sim t^{(3-D)/2} \qquad (4.32)$$

The dependence of the concentration of P, [P], on time is then given by

$$[P] \sim t^{(3-D)/2} \qquad (4.33)$$

Homonuclear Lindemann–Hinshelwood reactions

We start with the following homonuclear, bimolecular reaction

$$A + A \xrightarrow{k} C \qquad (4.34)$$

The product is an inert species C which takes no further part in the reaction scheme. You will be familiar with the following activation-controlled rate law which describes the change in the concentration of A, [A], with time

$$d[A]/dt = -k[A]^2 \qquad (4.35)$$

The solution of this equation is

$$[A] = \frac{1}{kt + 1/[A]_0} \qquad (4.36)$$

where $[A]_0$ is the value of [A] at $t = 0$. At long times $kt \gg 1/[A]_0$ so eqn 4.36 becomes

$$[A] \sim t^{-1} \qquad (4.37)$$

In the diffusion-controlled case, k is given by eqn 4.15. We know that in a time t a molecule A will sweep out a region of space with linear extent λ and volume of the order of λ^D. Eqn 4.25 tells that for $D_s < 2$, the time required to move through λ scales as $t^{1/(2+\theta)}$, which may be rewritten with the help of the identity 4.29 as

$$\lambda \sim t^{D_s/2D} \qquad (4.38)$$

When $D_s < 2$, the probability that the molecule of A will have visited every site in that volume approaches 1 as t becomes large, so that the number of particles in this volume can only be of the order of one. The concentration of A at a long time t is given by

$$[A] \sim 1/\lambda^D \qquad (4.39)$$

After substitution from eqn 4.38 we find

$$[A] \sim t^{-D_s/2} \qquad (4.40)$$

Thus, at long times we expect that the classical expression for the dependence of [A] on time (eqn 4.37) be substituted by eqn 4.40 when $D_s < 2$.

As the reaction progresses, the separation between the molecules that survive increases. When D_s is less than 2 the compact nature of the random walk means that molecules are less likely to diffuse together than would be the case for $E = 3$: [A] decays more slowly with time in eqn 4.40 compared with 4.37. When $E = 3$ the extensive nature of the random walk ensures a homogeneous distribution of reactants and we call this the *mean-field* case. The rate constant for the diffusion-limited process may now be calculated by substituting eqn 4.28 into eqn 4.15

$$k \sim t^{D_s/2-1} = k_1\, t^{D_s/2-1} \qquad (4.41)$$

where k_1 is a constant which is independent of time. The fact that the rate-constant k changes with time means that it is possible to prepare two reaction mixtures containing the same concentrations of the same reactants subjected to the same external conditions, but with quite different reaction rate constants on account of their different histories. Consider the two reaction vessels depicted schematically in Fig. 4.7, representing the same concentrations of the same reactants whose reaction rate is controlled by diffusion with $D_s = 4/3$. The upper vessel (Fig. 4.7(a)) contains a sample which had a much higher initial concentration, but which is being viewed at a later stage in its history than the lower (Fig. 4.7(b)). We fix the initial concentration so that at the time when this snapshot is taken, the concentrations of the reactants happen to be the same. Suppose that the times after reaction was initiated are 27 s and 1 s for (a) and (b) respectively. The ratio of the rate constants k_a and k_b is then given by eqn 4.41:

$$k_a/k_b = 27^{4/3-1}/1^{4/3-1} = 1/3 \qquad (4.42)$$

Thus, the reaction rate in Fig. 4.7(a) is three times slower than in Fig. 4.7(b). This is consistent with the way in which we would expect the system to evolve, with the survivors becoming more isolated with time. The distribution of molecules in the younger system favours much more rapid encounters.

If we substitute the expression (4.41) for k in eqn 4.36 and rearrange we find

$$[A]^{-1} = [A]_0^{-1} + k_1\, t^{D_s/2} \qquad (4.43)$$

At long times, when $[A]^{-1} \gg [A]_0^{-1}$ this may be approximated by

$$[A]^{-1} = k_1\, t^{D_s/2} \qquad (4.44)$$

Therefore

$$t \sim [A]^{-2/D_s} \qquad (4.45)$$

If we return to our basic rate equation (4.35) and substitute for k (eqn 4.41) we find

$$-d[A]/dt \sim t^{(D_s/2)-1}[A]^2 \qquad (4.46)$$

We may now eliminate terms in time by substitution from eqn 4.45:

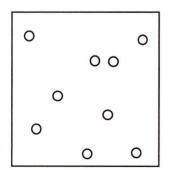

(a)

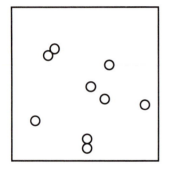

(b)

Fig. 4.7 Snapshot of the spatial distribution of reactant molecules in an homonuclear LH reaction with D_s = 4/3. (Sample with (a) high $[A]_0$ at t = 27s and (b) low $[A]_0$ at t = 1s. k_a/k_b ≈ 1/3.

$$-d[A]/dt \sim [A]^{1+2/D_S} \qquad (4.47)$$

Thus, for $D_s < 2$, the equation appears to have an anomalous molecularity despite the fact that the elementary rate equation has been assumed to involve just two molecules. If $D_s > 2$, $S(t) \approx t$ so that k is independent of time and the familiar rate equation (4.35) reappears.

Let us reflect on what we have uncovered. For diffusion-controlled reactions in an environment where $D_s < 2$, the rate constant k depends on time according to eqn 4.41. The decrease in k with t reflects the fact that as the reaction progresses, it becomes harder for the remaining reactant molecules to collide with one another for a given concentration. As a consequence, the reduction in [A] with the time and the molecularity of the rate both differ from what is expected for a conventional homonuclear bimolecular reaction where the reaction mixture is assumed to be homogeneous.

Heteronuclear Lindemann–Hinshelwood reactions

The kinetic behaviour of a LH reaction changes radically when we go from a homonuclear to a heteronuclear reaction. Consider the reaction

$$A + B \xrightarrow{\quad k \quad} C \qquad (4.48)$$

where C is an inert species with regard to further reaction under these conditions. The rate equation for the conventional, reaction-limited case is

$$-d[A]/dt = k[A][B] \qquad (4.49)$$

with a similar equation for the rate of disappearance of B. The general solution of this equation is

$$[A] = \Delta \left[1 - \frac{[A]_0}{[B]_0} \exp(-\Delta kt) \right]^{-1} \qquad (4.50)$$

where Δ is the difference between the initial concentrations of A and B i.e.

$$\Delta \equiv [A]_0 - [B]_0 \qquad (4.51)$$

When $\Delta = 0$, we find

$$[A] = (kt + 1/[A]_0)^{-1} \qquad (4.52)$$

and at long times, when $kt \gg 1/[A]_0$,

$$[A] \sim t^{-1} \qquad (4.53)$$

We may apply a similar form of dimensional argument to this case as we did to the homonuclear case. There is, however, a complication that arises through the inhomogeneity of the distribution of A and B molecules at the start of the reaction. If there are on average N_A molecules of A in a volume V, we expect a standard deviation of $\sqrt{N_A}$. Thus, in a region of linear extent λ, we expect to find ($[A]\lambda^D \pm \sqrt{[A]}\lambda^D$) particles. In a time $t \approx \lambda^D$ the particles in this domain will have had a chance to visit all the sites. Suppose there are more A molecules than B in this particular region. All the B molecules will be mopped up, leaving only those A molecules that were in excess. This will be of the order of $\sqrt{[A]}\lambda^D$ so that the concentration [A] in this region at that time is

$$[A] \sim \frac{\sqrt{[A]_0}\lambda^D}{\lambda^D} \tag{4.54}$$

$$\sim \lambda^{-D/2} \tag{4.55}$$

When $D_s < 2$ we may substitute for λ using eqn 4.25 which gives

$$[A] \sim t^{-D/4} \tag{4.56}$$

It has been shown that this relation is also valid for all physically meaningful dimensions and should therefore apply to diffusion-controlled reactions with $D = 3$.

Further reaction may only occur if molecules of A from a region that was originally rich in A diffuse into a region that started with an excess of B molecules. As the reaction proceeds, the inhomogeneities that arose from statistical fluctuations in the initial mixture become more pronounced and segregation into regions rich in either A or B occurs. This has a dramatic effect on the form of the rate equation. It becomes progressively harder for A and B molecules to find one another and the decay of either species with time is much slower than would occur with a mean-field reaction.

The apparently anomalous nature of the reaction kinetics arises because the concept of concentration that we usually use when we describe a reaction rate breaks down. The restrictions imposed by the requirement that reactants have to diffuse together means that only a small proportion of the reactants can 'get at' each other and reaction may only occur at the boundaries between the regions rich in either reactant. A pictorial illustration of this partition is given in Fig. 4.8.

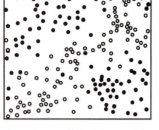

Fig. 4.8 Snapshot of the spatial distribution of molecules of A (open circles) and B (closed circles) when segregation of the reactants is apparent.

4.6 Observation of anomalous kinetic behaviour in fractal environments

A good experimental test of anomalous kinetic behaviour for an ER reaction is provided by an electrochemical process rather than a reaction between gas molecules and a surface. Suppose we dip an electrode into an unstirred solution of an ion or molecule that may react irreversibly at the electrode when a particular potential is applied. If the potential is applied suddenly at time $t = 0$, it will create a sudden depletion of the reactive species near the surface, and further reaction and hence the flow of current I will be controlled by the rate at which the reactive species can diffuse to the surface.

This in turn is controlled by the diffusion equation (2.31) whose solution with the appropriate boundary conditions predicts a dependence of I on t of the form:

$$I \sim t^{-1/2} \tag{4.57}$$

Suppose that the surface is not smooth, but has a fractal dimension $D \neq 2$. The rate at which a diffusion-limited reaction product forms under ER conditions at the surface may be obtained by differentiating eqn 4.33 with respect to t. This is equal to the current in the electrochemical process we are considering so

$$I \sim t^{(1-D)/2} \tag{4.58}$$

The verification of this expression is the subject of an elegant experiment in which a fractal electrode of known dimension was created by photolithography of a photoresist covering a gold surface. It was possible to create a gold electrode whose form was either a Sierpinski gasket ($D = \log 8/\log 3 \approx 1.893$) or a fractal carpet with $D = \log 3/\log 2 \approx 1.585$. The range of length-scales over which fractal scaling applied was of the order of 2 µm to 10 mm. The electrochemical reaction that was studied involved the electrodeposition of silver from silver nitrate solution; the decay of the diffusion-limited current after the potential of the electrode was changed did appear to obey eqn 4.58 as would be expected for the different values of D. We reproduce a set of results in Fig. 4.9.

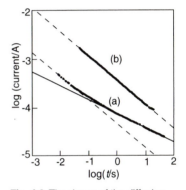

Fig. 4.9 The decay of the diffusion-limited current at (a) a fractal electrode with time t after the potential is changed to initiate annihilation of charge carriers at its surface. (b) Displays the response of a comparable flat, planar electrode.

Diffusion-limited homonuclear bimolecular reactions in an environment with $D_s < 2$ and diffusion-limited heteronuclear reactions in any environment have time-dependent rate constants and an anomalous dependence of reactant concentration with time when compared with mean-field reactions. We call such kinetic behaviour 'fractal', though it does not necessarily require a fractal environment.

The majority of chemical reactions that have been scrutinized for this type of fractal kinetic behaviour are spin-allowed photochemical or photophysical processes for which the reaction is very fast once the reactants have come together. Non-classical kinetic behaviour might then be observed for a heteronuclear bimolecular reaction or for a homonuclear bimolecular reaction for which the reactants are constrained to diffuse with $D_s < 2$. This latter case might be found if the diffusion is constrained to a tube in which D_s is effectively 1 or to the backbone of a percolating cluster.

In this context, the most extensively studied photochemical reaction is the photodimerization of naphthalene, represented by the following reaction scheme.

$$2N \xrightarrow{\text{hv}} 2N^* \longrightarrow N_2^{**} \longrightarrow N^{**} + N \xrightarrow{\text{hv'}} 2N \tag{4.59}$$

Naphthalene molecules N are excited from their ground states to a triplet N^* with laser light of frequency ν. These molecules may decay to their ground state, with the emission of green light. Alternatively, they may collide to produce a transient singlet dimer N_2^{**} which may decay to a naphthalene molecule in its ground state, and one in its first excited singlet, N^{**}. N^{**} decays to the ground state with the emission of ultraviolet light of energy $h\nu'$. The rate-determining step for the overall process is the diffusion-limited formation of N_2^{**} The concentration of the reactants N^* may be monitored through the level of green fluorescence. Under steady-state reaction conditions, which may be created by controlling the level of the stimulating light, the reaction rate is given by the instantaneous concentration of N^{**} and this may be monitored through the intensity of the ultraviolet fluorescence.

Such a reaction has been performed in several constrained environments in which diffusion occurs with $D_s < 2$. Naphthalene may be packed into microscopic tubes by crystallising it in a polycarbonate membrane which is 6 μm thick and contains cylindrical pores whose uniform diameters range from 10 to 1000 nm wide (Fig. 4.10(a)).

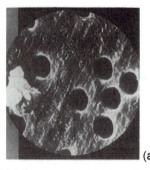 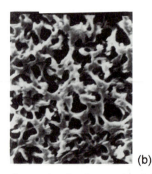

(a) (b)

Fig. 4.10 (a) Cross-section of polycarbonate membrane showing the regular distribution of pores (b) Electron micrograph of a porous nylon membrane (x10^4).

Reaction in a thin pore approximates to a diffusion-limited process with $D_s = 1$, and hence the exponent in eqn 4.41 is equal to $-1/2$ which is supported by experimental results. As the width of the pore increases relative to its length, D_s is observed to increase and the exponent tends to zero.

This photochemical reaction may also be performed in the less anisotropic porous membranes provided by a filter paper or the nylon membrane shown in Fig. 4.10(b). In these cases, the exponent of time in eqn 4.41 was found to be approximately 4/3 rather than the value of 0 expected for a mean-field process. If we interpret the anomalous exponent using a fracton diffusion model, we find this corresponds to $D_s = 4/3$ which is what we expect for diffusion on a percolating network.

Perhaps the best illustration of diffusion in or on a fractal concerns migration on a percolating network. Suppose we illuminate a crystal of naphthalene to create excited triplet states. The molecules are locked in place, but the electronic energy, called an exciton, may be transferred between molecules in a resonant process. If two excitons collide, they fuse to produce a singlet state and energy in the form of light in the UV part of the electromagnetic spectrum. Exciton transfer is much less efficient if it is between molecules with slightly different energy levels, as would be the case if one of them was the perdeuterated form $C_{10}D_8$. If we prepare crystals of naphthalene from a mixture of $C_{10}H_8$ and $C_{10}D_8$ we find that as the proportion of the deuterated material is increased, so the resonant transfer of excitons becomes less efficient until the percolation threshold is reached, when it plummets. The rate at which triplet fusion occurs may be followed by the rate of decay of fluorescence after the initial excitation of the system. Measurements on crystals of various compositions showed that in the region of p_c for this lattice, D_s took the value 4/3 as would be expected for a percolating cluster.

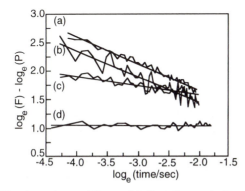

Fig. 4.11 The dependence of the annihilation rate constant on time for the photodimerization of naphthalene when the reaction is conducted in a porous membrane which restrains diffusion to narrow tubes. Curves (a), (b), (c), and (d) correspond to pores of radius 7.5 nm, 15 nm, 25 nm, and 400 nm respectively.

4.7 Summary

The interaction of molecules with surface fractals depends on whether the interaction leads to physisorption, chemisorption, or a reactive process. The surface available for physisorption may scale with the size of the probe molecule or, in the case of a powder, the size of the adsorbing particles. The scaling properties of a calibrated surface may be used to determine the effective area and hence orientation and conformation of other molecules.

Chemisorption and reactive encounters usually involve a subset of surface sites. The activity of a powdered catalyst may scale with r as r^{D-3}, where D commonly takes values that are smaller than those expected from physisorption measurements. D is sensitive to the specific combination of molecule and surface.

Diffusion in a fractal environment is characterised by the fracton dimension D_s which is defined as $D_s \equiv 2D/(2+\theta)$ where θ reflects the distribution of pathways for that particular object; $\theta > 0$ for fractals, and $= 0$ for Euclidean objects; D_s is equal to 4/3 for any percolating cluster. The rate at which a diffusing particle explores new sites is given by $S(t)$ which depends on time t as follows:

$$S(t) \sim t^{D_s/2} \qquad (D_s < 2) \qquad (4.60)$$

$$S(t) \sim t \qquad (D_s > 2) \qquad (4.61)$$

When a diffusion-controlled bimolecular reaction occurs in a fractal environment we may observe anomalous kinetic behaviour. Homonuclear annihilation reactions in $D_s < 2$ show time-dependent rate constants and molecularities different from 2. Heteronuclear annihilation shows similar behaviour in all physically meaningful dimensions and the reactants become segregated. As reaction proceeds, the concentration of particles that is available for reaction drops more quickly than for the mean-field case of a well-stirred reaction

I end with a cautionary comment. The visual and intellectual appeal of fractals has inspired very widespread application of the concepts I have presented. In some cases the quality of the data, or the range of length-scale over which the measurement is made does not justify the use of fractal arguments. Further, many of the materials or processes at which such analyses are directed are structurally, chemically, and electronically very complex and in order to treat them we have to make simplifications. When we observe a deviation from the predictions of Euclidean forms of such models, it is not necessarily clear that it arises from fractal behaviour or some other break-down in the model. The coverage of a rough surface with molecules of different radius may be influenced by small changes in the chemical character of the probe as well as by its size; the BET isotherm is based on questionable assumptions, so if it requires modification when applied to rough surfaces it may be for reasons other than fractal scaling; anomalies in kinetic behaviour may reflect competing reactions. We have already pointed out that the analysis of small-angle scattering data using fractal concepts often assumes some simplification such as monodispersivity of particles in a colloid. Nevertheless, when fractal concepts are applied to materials and processes at the appropriate length-scales, they may greatly enrich our understanding of regular and irregular solids.

Hamlet:	Do you see yonder cloud that's almost in shape of a camel?
Polonius:	By th'mass and 'tis, like a camel indeed.
Hamlet:	Methinks it is like a weasel.
Polonius:	It is backed like a weasel.
Hamlet:	Or like a whale.
Polonius:	Very like a whale.

Further reading

Chapters 4.1.2 and 4.2.1 in the book edited by Avnir contain information about adsorption of gas on surfaces, and Chapters 2.2.3, 4.1.1, and 4.1.3 describe diffusion and reactions in fractal environments. You may also find the following articles or reviews useful.

Avnir, D., Farin, D., and Pfeifer, P. (1983). Chemistry in noninteger dimensions between two and three. II. Fractal surfaces of adsorbents. J. *Am. Chem. Soc.* **79** 3558.

Farin, D. and Avnir, D. (1988). The reaction dimension in catalysis on dispersed metals. *J. Am. Chem. Soc*. **110** 2039.

Kopelman, R. (1988). Fractal reaction kinetics. *Science*. **241** 1620.

Rothschild, W.G. (1991). Fractals in heterogeneous catalysis. *Catalysis Reviews – Science and Engineering*. **31** 71.

Stauffer, D. and Aharony, A. (1992). *Introduction to percolation theory*. Taylor and Francis, New York.

Temporal fractals are treated in the following books

Gleick, J. (1993). *Chaos: making a new science*. Heinemann, London.

Schroeder, M.R. (1991). *Fractals, chaos and power laws: minutes from an infinite paradise*. W.H. Freeman, New York.

Fan, L.T., Neogi, D., and Yashima, M. (1991) Elementary introduction to spatial and temporal fractals. *Lecture Notes in Chemistry* **55**. Springer, Berlin.

Problems

4.1. The surface area of a fumed silica depends on the diameter d of the primary silica particles. The area A per gram of samples of the same silica whose particles are graded with selected values of d was estimated with BET (N_2) data, and is summarised in Table 4.5. Calculate the fractal dimension D for the surface of the particles over this range of length-scale, and estimate the maximum size of probe molecule that could be used to quantify D.

Table 4.5. BET (N_2) surface area A per gram of a fumed silica with different particle diameters d.

Sample	d (nm)	A (m^2 g^{-1})
OX50	40	49
130	16	123
150	14	141
200	12	186
300	7	265

4.2. Excitons E created by optical excitation in a crystal migrate over the lattice until they land on an impurity atom T that catalyses prompt radiative decay according to the following reaction scheme:

$$E + T \rightarrow T^* \rightarrow T + h\nu \qquad (4.62)$$

where T* is an excited state of T. Derive an expression for the probability that the exciton survives for a time t after creation, showing how it depends on the fracton dimension for the lattice. You may assume that this process is diffusion-limited.

4.3. An inorganic salt that is currently being investigated for use in X-ray films stores an image of the ionising radiation in the form of electrons and holes trapped at defects in the lattice. Those electrons and holes that lie close to each other may be prompted to recombine by a stimulated tunnelling process by exposing the salt to red light; this leads to emission of blue light. Further stimulation of luminescence is then limited by the rate at which new intimate pairs of electrons and holes form so that the luminescence intensity $I(t)$ observed at long times t under constant illumination exhibits slow decay and is proportional to $t^{-\alpha}$. Assume that both the electrons and holes are mobile and show how α relates to the dimension of a sample of the material that has been evaporated onto a smooth, inert substrate to produce a continuous solid whose fractal dimension D is less than 2 over a range of length-scales that include the diffusion length of the migrating species.

If the stimulating light is switched off at time t_1 and back on again at t_2, the signal is seen to recuperate; how is the degree of recuperation related to t_1, t_2, and D?

Answers to problems

1.1 380 m. **1.2** (i) $D = 2 \log3/\log5 \approx 1.365$ (ii) $2 \log3/\log7 \approx 1.129$ (iii) $2 \log3/\log3 = 2$. **1.3** Solution of the form $D = a^3/a^2$ where a is an integer; generator for one solution ($a = 2$) is shown in Fig. 4.12. **1.4** $D \approx 2.25$, σ (naphthalene) ≈ 55 Å^2 **1.5** See the section on chemisorption on page 71.

2.1 Subtract $u(\mathbf{r},n)$ form both sides of eqn 2.57. **2.2** See section 4.4 on pages 75 – 76. **2.3** Discussion in references in reading material; total length of all branches, $N(r)$, is proportional to r^D and $n(r)$ is proportional to $dN(r)/dr$ so $n(r) \sim r^{(D-1)}$.

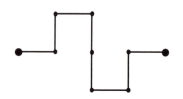

Fig. 4.12 Generator for problem 1.3.

3.1 Curve (a) is for M-5 which has surface dimension 2 and at longer length-scales is a mass fractal with $D = 1.9$; (b) is for HS-5 with surface dimension 2.2 and mass fractal dimension 1.8; (c) is for EH-5 with surface dimension 2.5 but dependence of $I(Q)$ on Q for longer length-scales is too indistinct for further analysis. Surface areas for M-5, HS-5 and EH-5 are 200, 325 and 390 m^{2}g^{-1} respectively. Primary particle diameter is about 90Å. **3.2** Lowest curve is for the (smooth) stabilised colloidal particles and the remaining curves are for samples with $W = 1, 2, 3$, and, 4 in ascending order on the graph. $W = 1$ gives very low branching and is a mass and surface fractal; the remainder are more highly connected and appear to be surface fractals only. **3.3** 3.65±0.02 ns; $D = 2.4$ and 2.7 at low and high concentration respectively, implying a smoother surface at shorter length-scales, but see P. Levitz and others in *Journal of Chemical Physics* (1991) **95** 6151.

4.1 $D \approx 2.02$ (smooth over this range of length-scales); particles range in size from $d_{max} = 40$ nm to $d_{min} = 7$ nm and $\sigma_{max} = (d_{max}/d_{min})^2\sigma(N_2) \approx 5$ nm^2. **4.2** Survival probability is proportional to $\exp(-[B]S(t))$, where $S(t)$ is given by eqns 4.60 or 4.61, according to D_s. **4.3** $I(t) \sim t^{-(1+D/4)}$; recuperation signal is proportional to $(t_1{}^{-(D/4)} - t_2{}^{-(D/4)})$.

Index